Additional Praise for *Here Comes the Sun*

"Trust Bill McKibben to find light in the darkness—and oh what a light it is! This is the energizing vision and game plan so many of us have been waiting for, and of course it comes from one of our era's most imaginative and trusted voices."
—Naomi Klein, author of
This Changes Everything: Capitalism vs. The Climate

"Bill McKibben has the unique ability to make important books both fun and a pleasure to read. For more than forty years he has campaigned about nature and our place in it, but *Here Comes the Sun* may be his most timely and necessary volley yet."
—Richard Cohen, author of
Chasing the Sun: The Epic Story of the Star That Gives Us Life

ALSO BY BILL McKIBBEN

The Flag, The Cross, and the Station Wagon

We Are Better Together

Falter: Has the Human Game Begun to Play Itself Out?

Radio Free Vermont

Oil and Honey: The Education of an Unlikely Activist

Eaarth: Making a Life on a Tough New Planet

The Bill McKibben Reader: Pieces from an Active Life

Fight Global Warming Now: The Handbook for Taking Action in Your Community

Deep Economy: The Wealth of Communities and the Durable Future

Wandering Home

Enough: Staying Human in an Engineered Age

Long Distance: Testing the Limits of Body and Spirit in a Year of Living Strenuously

Maybe One: A Case for Smaller Families

The Comforting Whirlwind: God, Job, and the Scale of Creation

Hundred Dollar Holiday: The Case for a More Joyful Christmas

Hope, Human and Wild: True Stories of Living Lightly on the Earth

The Age of Missing Information

The End of Nature

HERE COMES THE SUN

A Last Chance for the Climate and a Fresh Chance for Civilization

BILL McKIBBEN

W. W. Norton & Company

Independent Publishers Since 1923

Copyright © 2025 by Bill McKibben

Some material in this book originally appeared in
a different form in *The New Yorker*.

All rights reserved
Printed in the United States of America
First Edition

For information about permission to reproduce selections from this book,
write to Permissions, W. W. Norton & Company, Inc.,
500 Fifth Avenue, New York, NY 10110

For information about special discounts for bulk purchases, please contact
W. W. Norton Special Sales at specialsales@wwnorton.com or 800-233-4830

Manufacturing by Lakeside Book Company
Book design by Chris Welch
Production manager: Lauren Abbate

Library of Congress Cataloging-in-Publication Data is available.

ISBN 978-1-324-10623-4

W. W. Norton & Company, Inc., 500 Fifth Avenue, New York, NY 10110
www.wwnorton.com

W. W. Norton & Company Ltd., 15 Carlisle Street, London W1D 3BS

10 9 8 7 6 5 4 3 2 1

For Asa, of course.

May you grow strong and happy in the light of the sun.

And I say, it's all right.

—*George Harrison*

Contents

Introduction   1

SECTION ONE
WE COULD DO THIS

1   Notes on Burning   13
2   Notes on a Different Kind of Burning   42
3   Life on the S Curve   55

SECTION TWO
THERE'S NO REASON NOT TO DO THIS

4   Can We Afford It?   81
5   But Can the *Poor* World Afford It?   101
6   But Is There Enough Stuff?   121
7   Do We Have Enough Land?   137

SECTION THREE
LET'S DO THIS!

8	Time to Push	157
9	A Subtly New World	180
10	Face to the Sun	190

Notes on Sources and Acknowledgments   209

Introduction

As I wrote the first words of this book, late in December 2024, a warm rain pelted the Green Mountains of Vermont. It should have been snow, of course, but scientists had just announced that 2024 was the hottest year ever recorded—indeed, the paleoclimatologists said it had been the hottest year in the last 125,000. While I was still immersed in typing these pages, America inaugurated Donald Trump as president after he ran on the premise that global warming was a hoax. Oh, and Los Angeles caught fire; among the least important results of that conflagration was that the home of my earliest memories, up against the Altadena hills, burned to the ground.

If I have a literary reputation, it's for a kind of dark realism. When I was still in my 20s—way back in the 1980s—I published what is sometimes called the first book on the climate crisis. It bore the cheerful title *The End of Nature*; in the decades since, with 20 books and countless essays and articles published, I have chronicled those early warnings as they came true. This moment would seem to be—indeed it is—the summation and the vindication of all that angst.

And yet, right now, really for the first time, I can see a path forward. A path lit by the sun.

And it's a path not just out of the climate crisis—it's a path that opens into a very new world. As I type, I've got this book's titular song, George Harrison's gentle and optimistic anthem, pouring through the headphones, blotting out the sound of the rain on the roof. I think that even as we teeter on the brink of renewed fascism, we're also potentially on the edge of one of those rare and enormous transformations in human history—something akin to the moment a few hundred years ago when we learned to burn coal and gas and oil, triggering the Industrial Revolution and hence modernity. But now, quite suddenly, we're learning *not* to burn those fossil fuels, and to rely instead on the large ball of flaming gas that hangs 93 million miles distant in the sky. We're on the verge of realizing that the sun, which already provides us light and warmth and photosynthesis, is also willing to provide us the power we need to run our lives. We are on the verge of turning to the heavens for energy instead of to hell.

It won't happen automatically, and I don't know if we *will* do it, at least in the short window physics is giving us to deal with climate change. In his first hours in office Trump did all he could think of to steer America and the world away from this bright future; the main headline in *The New York Times* on his second day was "Trump Wants to Unleash Energy, as Long as It's Not Wind or Solar." So no, it won't be easy. But I am convinced it *could* happen, and certain that it should. In a world where almost everything seems to be going wrong, this is the one big thing suddenly going right. I am willing to fight for it, and I hope you will be, too.

In one sense, this book is very time-bound. Sometime in the early part of the 2020s we crossed an invisible line where the cost of producing energy from the sun dropped below the cost of fossil fuel. That's not yet common knowledge—we still think of photovoltaic panels and wind turbines as "alternative energy," as if they were the Whole Foods of power, nice but pricey. In fact—and more so with each passing month—they are the Costco of energy, inexpensive and available in bulk. We live on an earth where the cheapest way to produce power is to point a sheet of glass at the sun; the second-cheapest is to let the breeze created by the sun's heating turn the blade of a wind turbine. Beginning about the middle of 2023, we entered the really steep part of this growth curve that could redefine our future, crossing another invisible line, this one marking the installation of a gigawatt's worth of solar panels on this planet every day. (A gigawatt is about the output of a typical coal-fired power plant or nuclear reactor.) By the fall of 2024 that gigawatt was going up every 18 hours. We're still in the early days of this transformation—right now only about 15 percent of the planet's electricity comes from sun and wind, and only about a quarter of the energy we use comes from electricity. But exponential growth changes numbers like that very fast—in 2024, 92.5 percent of all new electricity bought online around the world came from renewables; in the US the figure was 96 percent. By April 2025, fossil fuel was producing less than half of American electricity, for the first time ever. There's no longer a technical or financial obstacle in the way; we already have the factory capacity, mostly in China, to produce as many solar panels as the climate scientists say we need. In May 2025 came the news that China had used 5 percent less coal in the first quarter of the year to produce electricity than it had in 2024—despite a surging economy, Chinese emissions were actually dropping.

The suddenness of this moment is startling. The solar cell was invented in 1954, and it took from then until 2022 to install the first terawatt worth of solar power on this planet. It took two years to get the second; the third will be quicker still. *It's all brand new.*

But there are a few places that are running far ahead, showing what's possible. China is well on its way to being the earth's first "electro-state"; something like half of all clean energy has been installed within its borders. And 2024 was a breakout year in California: there were finally enough solar panels that for parts of most days the state could produce from renewable sources more than 100 percent of the electricity it used; at night great batteries that had spent the afternoon soaking up sunshine often became the biggest source of supply to the electric grid of the world's fifth-largest economy. As a result, in 2024 California used 25 percent less natural gas to produce its power than it had in 2023, which is a big number. Through mid-April of 2025, as more panels and batteries came online, the numbers got even better: California was using 44 percent less natural gas to make electricity than it had just two years earlier. On the other side of the world, in Pakistan, a flood of cheap solar panels from China let homeowners and storekeepers and factory managers build the equivalent of a third of the country's electric grid inside of a year. Peasant farmers, often just laying the panels on the ground, started pumping their irrigation water with electricity instead of generators powered with fossil fuels; diesel sales dropped 30 percent in the course of a year.

Those kind of shifts, replicated quickly in many more places, could take a real bite out of the grim predictions of climate scientists; the sun burns so we don't need to. We are in a desperate race; those scientists have told us that to stay on anything like a survivable path we must cut greenhouse gas emissions in half before the decade is out. That target is on the bleeding edge of the technically possible, and this book is an effort to shove us toward that deadline.

But I hope that this book is timeless as well—that it's anticipating a shift that will play out over many lifetimes, and in ways that diverge dramatically from our recent history. That's because energy from the sun is not just cheap. It's also *diffuse*, available everywhere instead of concentrated in a few places. And that prefigures a different world with a more localized and more humane geopolitics; indeed, the sun works more reliably toward the equator, which could allow the redress of some of earth's great inequities. In February 2025 the energy analysts at the Rocky Mountain Institute reported that renewable energy was growing twice as fast in the developing world of the Global South as in the developed world of the Global North. Relying on energy sources that are abundant instead of scarce—the sun and the wind each day produce thousands of times as much energy as we could ever use—could even reconfigure our ideas of competition and conquest. Unlike oil and gas, sun and wind can't be hoarded. If fascism scares you the way it does me, figuring out how to break the centralized power of the fossil fuel industry is a key form of resistance.

And for a species that has become almost fatally disconnected from the natural world, the sun offers a way back into a relationship with reality. We were all sun worshippers once; it's not perhaps too much to imagine that we might someday soon gaze up a little more often, maybe even breaking a little of the enchantment woven by the glowing lights in our palms. This is not, I think, a "technofix," but something far more fundamental. We have the chance to join in a great global project, providing affordable energy to every human community even as we stave off our greatest threat. It could prove a unifying mission for a divided world. The last remotely comparable project was the moon shot of the 1960s, but that involved one nation putting two men on an orbiting rock. This quest involves bringing our star down to earth to make that earth work—what could be more quintessentially human?

All this hope risks sounding giddy; let my dark realism reassert itself for a moment and offer up some caveats and cautions. I'm not overly concerned about the things people usually point to. As I'll make clear, we're not going to run short of minerals to build batteries or land to put panels on. Instead, my worries stem from hard realities both physical and political.

First, this definitely comes too late to "stop global warming." We've already done fundamental damage to the planet's physical systems, to the point of altering the jet stream and weakening the Gulf Stream; we've already raced past the 1.5 degree Celsius rise in global temperature that we pledged in Paris to avoid. (In April 2025, the Trump administration fired most of the American scientists who monitor this increase, perhaps reasoning that what we don't know can't hurt us.)

Our best hope now is simply to stop the heating of the earth short of the point where it cuts civilization off at the knees, and even that will be a very close call. I will return to the question of pace over and over in these pages, because it's what matters most. I have little doubt we will run the world on sun and wind 40 years from now, but if it takes us anything like 40 years to get there then it will be a broken planet; our energy sources will hardly matter. The march of history won't get us where we need to go fast enough; we need to force that march.

Second, there's no guarantee that the momentum of the last few years will continue. The fossil fuel industry has read the numbers too, and so they've girded for the fight. As the chairman of one big oil company said in the fall of 2024, the industry thinks we should keep burning gas and oil until "every last molecule" had been sucked from the earth. If you think that capitalism guarantees we'll pick the lowest-priced option, think again: In certain ways, solar and wind power are almost too cheap for our economy. Investors who have gotten rich controlling the hoarded "reserves" of fossil fuel are scared of the fact that the sun delivers energy for free each time it rises above the horizon, and in their fear they're massively gaming our political system. The worldwide elections of 2024 saw setback after setback, with oil-soaked populists winning control in too many places. Just as they played the game of climate denial with real success for three decades, they now engage in a kind of solutions denial, claiming we're not ready for clean energy, or offering up substitutes closer to the status quo. Some of these substitutes (geothermal power and nuclear energy, if the cost ever comes down) may offer useful side dishes to the main course of sun, wind, and batteries; others (carbon capture from power plants, biofuels) are just expensive efforts to extend the business model of this industry a little longer. All of the substitutes are effective at distracting us, especially in the distorted infosphere of greenwash and spin we inhabit.

Nowhere, of course, is that distortion more powerful than the United States, where Trump rode back into office vowing to "drill, baby, drill" and to crash the electric vehicle (EV) industry. He'd been in office four hours when he signed an order ending all federal support for wind power. (As for solar energy, the week before the election he said, "It's all steel and glass and wires. It looks like hell. And you see rabbits get caught in it.") By April, just three months into his second term, Trump was announcing plans to revive the coal industry, and his bizarre tariffs were making life harder for renewable energy developers; he cut off funding to Princeton's climate modelers on the grounds that their findings were causing "climate anxiety." All of which is to offer a third caution: Just because the world goes in one direction, that doesn't mean every nation will follow. Yes, there's enormous momentum behind this transformation; on the last day of February 2025 the federal Energy Information Administration predicted that 93 percent of American electric generation built in Trump's first year would be carbon-free, mostly from solar. In the first month of 2025, as Trump was taking office, sun and wind combined made up 98 percent of new generating capacity in the States. But clearly the Trump/Musk team will try to break that momentum; already-high tariffs on Chinese solar panels are being increased again even as I finish this manuscript, and the administration is embarked on a sprawling effort to achieve "energy dominance" based on oil and gas. It's an effort to stuff the solar genie back in the barrel, and we don't know yet to what degree it will succeed. The Biden administration, with the Inflation Reduction Act, set in motion transformative spending on clean energy technology, and spread the money carefully around the red states; Texas, home base of the hydrocarbon industry, is now outpacing even California in clean energy (though the state legislature, as of spring 2025, was engaged in an all-out effort to sabotage that growth). Power from the sun can appeal to conservatives ("my home is my well-wired castle") as powerfully as it does to liberals. But the addiction to fossil fuels and all its accomplices (the giant SUV, say) runs deeper here than anyplace else; it will be a fight to turn the American page.

 I'm ready for that fight. Even as I write these pages, I'm helping organize what we're calling Sun Day, set for the autumnal equinox in September 2025. Indeed, some of the proceeds from this book are supporting that organizing process, because its goal is the same: to help people understand the possibility of our moment. As we shall see, much of the progress that engineers have made has come on the back of inspired activism, something we need more of. In this fight, the solar panel and the wind turbine are both the crucial machines and also the symbols of potential liberation.

And in true Hollywood fashion, our liberation and our destruction are arriving at precisely the same time, offering us a remarkable choice. Everything is going wrong, except this one big thing. Our species, at what feels like a very dark moment, can take a giant leap into the light. Of the sun.

SECTION ONE

WE COULD DO THIS

1

Notes on Burning

We're quite suddenly at the moment where, as a species, we could and should break the habit of burning things. We could and should extinguish the fires in our power plants and factories, beneath the hoods of our cars, in our basements and kitchens, replacing those blazes with the fire of the sun. That's the revolutionary idea of this book: With one large exception, which I'll get to eventually, we could leave fire behind. We don't really need it anymore.

But we sure needed it in the past; indeed, understanding the significance of this moment means understanding the significance of fire to human history. The eclectic and graceful historian Alfred Crosby once pointed out in his classic book *Children of the Sun* that the earth is "unique among the planets of our solar system in its inclination to catch fire." Venus has volcanoes, and Jupiter has lightning, but those aren't fire—only organic matter can burn. Because the earth has photosynthesis, we have lots of grass and wood, and because photosynthesis gives off oxygen we have lots of that too, and that's all you need for fire.

As far as we know, the earth is also unique (Crosby again) "in having inhabitants to ignite and control, to domesticate, fire." Almost certainly early humans—*Homo erectus*, probably—

encountered fire in the wild, as lightning touched off forests. (Chimpanzees will follow fires, collecting charred animals to eat.) At some point those ancestors learned to control flame, and eventually to produce it, and with that it became possible to move away from the equator, and to cast light deep into caves, to harden spear points and to burn off grasslands. Above all it became possible to cook—applying fire to food means that some of the processes of digestion can be performed outside the body. This ability is reflected in our very form—it's why we have relatively small guts and teeth. Cooking made meat much easier to digest, giving us a calorie-dense diet that supported the evolution of a much-larger brain. (This is why celebrities on raw-foods diets lose weight.) Fire also kills bacteria and detoxifies some poisons in plants; we even learned to smoke and thus preserve food and to survive on staple grains—wheat, rice—that can be stored for long periods but need to be heated. I'll quote Crosby once more: "Cooking is universal among our species. No explorer ever found a human society that did not cook. . . . Animals do at least bark, roar, chirp, sending signals by sound; only we bake, roast, and fry."

When exactly we started cooking food is hard to figure out; some anthropologists work backward from the human body, figuring that the small teeth are a sure sign, while others scour caves for fire rings and then try to date them. At Wonderwerk Cave in South Africa, the first microscopic traces of wood ash and animal bones located deep enough in the cave that they couldn't have been caused by a lightning strike come from a layer about a million years old. Over the eons that followed, some anthropologists posit that standing around the campfire night after night talking over the events of the day helped develop the social bonds that mark our species—a kind of proto-Zoom. (Darwin contended that fire and language are the two things that set us apart, and in that sense those are connected.)

Eventually we learned to use fire for many things—smelting metals, for instance. Mostly we burned wood; peat, which is halfway between wood and fossil fuel, probably came next. The earliest known use of coal is for carved ornaments, but eventually people learned it burned too, perhaps beginning in China. The British began to use coal at a serious pace in about the 13th century—at first it was called sea coal, because most of it was collected from the easy-to-find outcrops along the island's beaches. Before long the British were mining coal, and in enough quantity that by 1661 there was a book, conceivably the first environmental tract, titled *Fumifugium: Or, The Inconveniencie of the Aer and Smoak of London Dissipated*. (Inconvenient truth indeed.) The crucial developments, however, awaited the next century.

In the fall of 2021, I made my first post-Covid overseas trip, to the global climate conference which was held that year in Glasgow, Scotland. I was scheduled to give a talk in the chapel of the University of Glasgow, and, arriving early, I wandered the grounds. (They looked familiar—as it turns out, it's where they shot the exteriors for the Harry Potter movies.) I wandered into the Hunterian Museum, and there found a working model of the world's first steam engine, built by one Thomas Newcomen in 1712. It was a lovely machine, all gears and wheels and boilers, and it burned coal, generating enough force to pump the standing water out of the bottom of coal mines, thus making it easier to . . . get more coal. Call this the John-the-Baptist engine, prefiguring what was soon to come; in one of those lucky historical twists, a man named James Watt was employed as an instrument maker at the University of Glasgow, and in the latter part of that century he was summoned to make repairs on this very machine. As he worked, Watt realized that with a few simple tweaks—

essentially, a separate condenser—he could double the power of the engine. Which was enough to change everything.

The great economist John Maynard Keynes once noted that for most of human history—from, "say, two thousand years before Christ down to the beginning of the eighteenth century"—there "was really no great change in the standard of living of the average man in the civilized centers of the earth." At best, he said, our standard of living had doubled over that long term. And that's because we didn't learn to do much of anything new. When history began we already knew about the wheel and the plow and the sail and the pot. The prime mover was muscle, human or animal; if you needed a lot of energy, you could harness some oxen. (Cooks spun rotisseries by having dogs walk on treadmills.) For all those millennia we essentially depended on sunlight transmuted into calories transmuted into arms and legs. But now, thanks to Watt and his machine, we could efficiently take ancient sunlight stored in the form of coal and use it for anything: powering trains along tracks far faster than horses, propelling boats across oceans more swiftly than wind, pushing looms, and spinning lathes. As the economist William Stanley Jevons wrote in 1865, "Coal in truth stands not beside but entirely above all other commodities. It is the material energy of the country—the universal aid—the factor in everything we do." Without it, he wrote, "we are thrown back into the laborious poverty of early times," by which he meant every time before Watt.

In fairly quick succession humans also figured out about petroleum, which in certain ways improved on coal—oil provides half again as much energy per pound, and it's easier to store and to transport. When the first modern oil well was drilled in Titusville, Pennsylvania, in 1859, the very prescient might have begun to sense the geopolitical balance shift across the Atlantic. John D. Rockefeller formed Standard Oil in 1867, and before long controlled most of the refining and pipeline capacity in America. This was a remarkably profitable monopoly, because at about the same time the internal combustion engine was invented in Germany—lighter and more adaptable than the steam engine, it of course reached its apotheosis under the hood of that other new invention, the automobile. Natural gas, the third of this hydrocarbon trinity, was another 18th-century wonder, especially after Robert Bunsen (of burner fame) figured out how to produce a continuous flame.

One way to explain the impact of these inventions is that humans (in the Western world, anyway) had been freed from photosynthesis; another was that we'd each acquired an army of servants that worked for very little. A barrel of oil contains the energy equivalent of 25,000 hours of human labor—a dozen years of 40-hour weeks. It's as if, the geographer Samuel Miller McDonald wrote, "humans had figured out how to send loggers, hunters, and farmers back through time to carve up, shovel out, and burn those sedimentary layers of life." Now, instead of it taking 4,000 years to double the standard of living, it could and did happen in decades, but only if you were in the lucky places. England's new coal-fired mills more or less killed off India's textile industry, for instance. As Crosby notes, before the Industrial Revolution, India, China, and Europe had accounted for 70 percent of the world's GDP, and in roughly equal shares. By 1900 Europe was producing 60 percent of the world's stuff, China 7 percent, and India 2 percent. Europe was also producing vast amounts of pollution, but who cared? The Yale historian Sunil Amrith quotes a Sheffield "smoke inspector" as saying no one would want a smokeless England because despite its "purer air, clear skies and more sunshine it would be a country of universal poverty." One can read Dickens or Engels for descriptions of the murk and danger of those factory towns, but the material gain that came from all that energy eventually submerged most protest.

By the middle of the 20th century, everyone was in broad agreement: Combustion was the central fact of our world. Amrith quotes a Soviet propaganda poster: "From oil we take for the needs of our country a river of gasoline." Mao set his people to work building backyard blast furnaces to catch up with the steel production of the superpowers. Nehru turned his back on Gandhian home industry and started building factories. Brazil discovered oil reserves in the 1940s, and it became a source of great national pride—the national oil company Petrobras was for a time the biggest company in the Southern Hemisphere. (I remember, in the 1980s, visiting a small pilot refinery in the hinterlands of southern Brazil, where they were producing some version of synthetic crude. A few days before, the manager told me, a nationalist politician had visited, and had insisted against all advice in washing his hands in the oil of the motherland; apparently the ceremonial dinner that had been set to follow was canceled when soap couldn't remove the stains.) In 1900 the world's GDP had been about a trillion dollars; by 2000 it was 30 times that. Our world had become a machine for turning fossil fuel into wealth.

No country, of course, figured out how to use this new power more effectively than the United States. The greatest spurt of wealth in human history followed America's victory in the Second World War, and most of it depended on that tiny flame from the spark plug. We built a whole new kind of place—the sprawling suburb, defined by endless cul-de-sacs and connected by a network of roads that eventually fed into the new interstates. Life in these places was impossible without a car, but we *loved* cars. We learned to eat in them, to watch movies from them. (A certain form of romantic encounter was literally called "parking.") This was the world where I came of age, at the wheel of my family's maroon Plymouth Fury. By 1970, as I approached teenagerhood, America consumed a third of the world's energy. Our national project was mostly building bigger houses farther apart from each other and then driving between them.

You can argue whether or not all this fire produced happiness, but you can't argue against its centrality. When upheaval in the Middle East threatened our supplies of cheap oil, the resulting chaos elected Ronald Reagan. A few years ago I had lunch with one of his successors, and the first words he said were "The most salient fact in American political life is the price of gas." Not far from my house, during the last presidential campaign, a Trump supporter put up a sign that said, "I'll trade some mean tweets for $1.89 gas."

And not surprisingly, almost everyone else followed us down this path. The world has burned more fossil fuel with each passing year, right through 2024. Our planet is now a series of endless controlled blazes, and the sum of all those fires has made our species a geological force. Before the Industrial Revolution, volcanoes were the main source of carbon dioxide in the atmosphere; now we produce 60 times more CO_2. (There are single American states that emit more than all the active volcanoes in the world each year.) Fire defines us.

And of course it also threatens us. Worldwide, nine million people die each year—one death in five—from breathing the particulates that come with combustion. And now all that carbon dioxide from all those fires is breaking the climate system, and with it all that we hold dear.

So many threads of this story came together for me in that Glasgow museum. Just around the corner from the Newcomen engine, I happened on another exhibit, this one devoted to the chemist Joseph Black—university technician James Watt had helped fix his equipment too, which is ironic, since it was Black who discovered carbon dioxide. There was also a marble bust of Adam Smith, another Glaswegian of the same vintage. I've always thought it right that he published *The Wealth of Nations*, the founding document of modern capitalism, seven years after Watt had built his engine; one was not possible without the other.

But in 2021, at the climate conference, all the various contradictions that stemmed from those discoveries and inventions were coming into focus. I gave my talk in the chapel and then walked downtown to meet my friend Reverend Lennox Yearwood, one of the world's best climate organizers. He was wearing a baseball cap, and across the brim it had the legend Nine Years. That's because 2030 is the deadline that the world's climate scientists had given us for cutting emissions in half. The history of the Industrial Revolution convinces me that we won't make that goal by cutting *our use of energy* in half—I doubt that we will, at least in significant numbers, change our behavior speedily enough to matter. Instead, we'll have to find another less dangerous way to slake our thirst for energy.

Which is why it's nice that I have one more history to recount, one that puts combustion on the . . . back burner.

This new story begins—well, you could argue that it begins with the Greeks and Romans, who used mirrors to focus sunlight on attacking naval vessels. The obvious power of the sun interested inventors, and also cranks, from the start. As Richard Cohen recounts, in his epic history *Chasing the Sun*, Leonardo da Vinci proposed a mirror four miles wide, which would have taken more glass than existed at the time. Antoine Lavoisier, who gave oxygen its name and discovered hydrogen, also built a "solar furnace," which looks like a giant magnifying glass. Lavoisier's career was interrupted by the guillotine, but in the next century another Frenchman, Augustin Mouchot, had constructed a steam engine powered by the sun—an inverted "mammoth lampshade" that could power a printing press. Cohen describes a Baltimore inventor, Clarence Kemp, who patented the first commercial sun-powered water heater in 1891; he went on to build a 15-horsepower solar pump that required 1,788 mirrors and was used to irrigate a Pasadena ostrich farm. Jonathan Swift had pre-parodied such efforts in *Gulliver's Travels*, where the first scientist he meets on the levitating island of Laputa had been working for eight years to "extract sunbeams out of cucumbers, which were to be put into vials hermetically sealed, and let out to warm the air in raw inclement summers."

Even when they worked, these early efforts to capture the direct power of the sun were overwhelmed by the cheap and easy power flooding in from coal and gas and oil. By 1940 George Gamow, the theoretical physicist, wrote that "the direct utilization of solar heat . . . is employed only in a few tricky devices—to run the refrigerators of cold-drink stands in the Arizona desert, or to heat waters for the public baths of the Oriental city of Tashkent."

If sunlight was really going to power the modern world, it would need to come via a very different route. Back in 19th century France, Edmond Becquerel, a 19-year-old working in his father's laboratory had discovered that shining sunlight on to plates of platinum, silver, or gold produced an electric current. Other engineers tinkered with this "photovoltaic effect" for the next century, but it wasn't until 1954 that it finally took modern form. That happened at Bell Labs in New Jersey, which was perhaps the most emblematic locus of American power in the great postwar years. Funded by Americans making phone calls, the lab was a kind of Gulliverian paradise of pure and applied research;

it pioneered radio astronomy (and found the background cosmic radiation that confirmed the big bang theory); it produced the first laser and the first transistor (and also the original vocoder!). And on an early spring day in 1954 a trio of researchers, who had been working on power sources for remote telephones, announced the first practical photovoltaic cell, a silicon-based device that managed to convert about 6 percent of the sunlight that fell on it into usable energy. That was enough for the Bell "Solar Battery" to power a tiny toy Ferris wheel and a small radio transmitter, and in the process shift the world.

The New York Times that day featured stories about a new US effort to "end the war in Indo-China," and about plans for 28 eating places along the brand-new New York State Thruway, ranging from "snack bars to swanky cafes." The US announced that the Iraqi government would receive "military help," bringing to "29 the number of countries receiving aid under the United States mutual defense assistance program," and a minister gave a farewell sermon to his downtown congregation, warning that too many Americans felt as if they were "live pawns in a chess game over which they have no control." Philco advertised its 1954 line of air conditioners, available at Wanamaker's ("mount flush with window sill!"); and the newly crowned Mrs. America, a "tall St. Louis blonde" named Mrs. Madison Jennings who possessed "a 35-inch bust, a 25-inch waist, and 36-inch hips," insisted in her victory statement that "I'm still a housewife," and "to prove it headed for the kitchen of the Mrs. America cottage in Daytona Beach to prepare lunch for her husband and son."

And there, on the bottom of page one (right next to a story about the beginning of the widespread polio vaccination the next day), there was an account of the Bell Labs press conference, under the headline "Vast Power of the Sun is Tapped by Battery Using Sand Ingredient." The new device, *The New York*

Times reported, was a "simple-looking apparatus made of strips of silicon, a principal ingredient of common sand. It may mark the beginning of a new era, leading eventually to the realization of one of mankind's most cherished dreams—the harvesting of the almost limitless power of the sun for the uses of civilization." The sun, the article noted, "pours out daily more than a quadrillion kilowatt hours of energy, greater than the energy contained in all the reserves of coal, oil, natural gas and uranium in the earth's crust."

Let us pause for a moment just to understand the workings of this small piece of silicon that serves as the interface between the sun and the things we need: warm homes, cold beer, a way to move easily around the world. As *The Economist* magazine put it recently, "A photovoltaic cell is a very simple thing: a square piece of silicon typically 182 millimeters on each side and about a fifth of a millimeter thick, with thin wires on the front and an electrical contact on the back. Shine light on it and an electronic potential—a voltage—will build up across the silicon. Run a circuit between the front and the back, and in direct sunlight that potential can provide about seven watts of electric power." Photons from sunlight knock loose electrons from the silicon; that's what creates the power: a tiny reaction, which needs to be magnified endlessly.

You'll recall that it took the better part of the 18th century for James Watt to double the power output of Thomas Newcomen's original steam engine. It's been 70 years since that Bell Labs press conference. In the time since, the efficiency of a solar cell has quadrupled, from 6 percent to about 25 percent. (Some experimental cells are nearing 40 percent efficiency.) The world (and by the world, I mean mostly China) now produces 70 billion of these solar cells a year, sandwiching them between glass to produce the panels that end up on rooftops or in solar farms.

We're very lucky that the sun's temperature, in the words of Oxford physicist Keith Barnham, "is such that the band-gap energy of silicon, one of the most abundant elements on earth, is nearly ideal for converting the energy in sunlight to electricity efficiently." And we're lucky that solar panels work best when we need them the most, during the middle of the day when we're at work.

We're lucky, too, that the sun heats the world to different temperatures in different places, because that produces the wind that is the second most important source of solar power. This technology started to come of age at pretty much the same time as the solar cell: Though humans have long used windmills to pump water and grind grain, the world's first megawatt-sized turbine came online in the early 1940s, a little ways south of my home in the Vermont town of Castleton. An MIT grad named Palmer Putnam (who would later serve as the president of his family's publishing company) convinced the local utility that the breeze sweeping a Green Mountains promontory called Grandpa's Knob could be successfully harnessed. Vannevar Bush, who ran the nation's scientific research during World War II, approved Putnam's design for his turbine, which had two blades, each 66 feet long and weighing eight tons. The contraption ran from October 1941 to February 1943, when a shaft bearing failed; the war meant that no one could scrounge the part until 1945 when the machine restarted, only to lose a blade three weeks later.

Wind power is not quite as elegant as the electricity a solar panel produces—you need a moving part, the turbine blade, to rotate coils of wire in a magnetic field, producing energy in just the way that Michael Faraday first described in the early 1800s. But it's graceful in another sense: Because it takes awhile for the sun to heat the air molecules that produce the wind, wind tends to build in power later in the afternoon, just as the photovoltaic effect begins to ebb. Not only that, but the farther north you go, the stronger the wind gets—which is useful, since Norway has rather less sunlight than, say, Greece. And wind speeds tend to be higher in the winter than the summer, because of sharper temperature gradients.

Sun and wind are therefore complementary. And they have a triplet, hydropower, which is just one more form of solar energy. After all, sunlight falling on the ocean increases its heat energy, till some water molecules have enough energy of motion to become the water vapor that eventually forms clouds, and then to fall as rain in the mountains where they can be stored behind dams and used, like wind, to turn a dynamo. Hydropower is a mature energy source; we don't need more big dams (indeed it probably makes sense to tear some down), so I won't be writing much about it in these pages. But those dams provide reliable baseload power. And together with the fast-expanding power of photovoltaic cells and wind turbines, hydropower provides the possibility of a world that runs without combustion. A world where we can put out the fires that plague us without extinguishing the economy those fires produced. A world that might just work.

Sun and wind promise a new flood of electricity. But for that flood to put out the fires of global warming, we have to do more than just put up lots of panels and turbines. We also have to convert our economy to use all that electricity. We have to switch, that is, from an energy system based on heating things up to one geared to producing what scientists call "work," which is essentially making things move. This is no small shift: Wood and coal and gas and oil all store energy in molecular bonds; the heat of a fire breaks those bonds, producing more heat. So, for instance, a campfire produces the warmth you want to ward off the night's chill and to cook food. In a conventional car engine, a spark plug ignites a mixture of gas and air, producing an explosion that moves a piston, turns a crankshaft, and powers the wheels—the burned gases exit through the exhaust pipe. Three centuries after Thomas Newcomen and James Watt, most of our energy comes from heat.

But the sun, as we've seen, exerts its force by moving electrons, and hence can, say, turn a magnetic rotor in an engine. This, as it happens, is a far more efficient way of getting things done. As a team from the Rocky Mountain Institute explained in a report in the autumn of 2024: "Burning gas to light a room creates more heat than light. Burning coal to create electricity creates more heat than electricity. Burning oil to move a vehicle creates more heat than motion. We are sending more energy up smokestacks and out exhaust pipes than we are putting to work to power our economy." This is not hyperbole: Burning oil to power a Ford or burning coal to produce electricity is at best just over 30 percent efficient—or, 70 percent inefficient. (And for reasons of thermodynamics, that's about as good as it's ever going to get.) So you *can* use heat to get things done—our whole energy system is at the moment geared around doing just that—but you wouldn't want to if you had a choice, which we now do. As the RMI team put it, "For over a century we've been using the hammers of heat supply to bash in the screws of work demand. We got quite good at it and made it work—as long as you ignore the incredible inefficiency of the botched job. Now, with renewable electricity, we've found a screwdriver."

When you start thinking about this, it stretches your mind. We currently take for granted an exceptionally complicated system, and we think of the far simpler possibility as "alternative energy." Consider a car powered by an old-fashioned internal combustion engine. For it to work, someone has to find oil, an increasingly difficult task (at a new field, they're likely to be drilling a mile or two beneath the surface of the ocean). That liquid has to transported to a refinery, where it is distilled, cracked, and then blended (a refinery can cover hundreds of football fields with its complicated array of pipes and tanks). It's then shipped to a gas station, where you fill the tank of your car (making sure you don't create a spark while you do so). The drivetrain in an internal combustion engine can have 2,000 parts—it's pretty complicated to control a series of explosions and cool the resulting heat. By contrast, my EV—a Kia Niro—sits in the garage about 25 feet from the solar array on my roof. The electricity those panels produce run through a cord into the motor, which has about 20 moving parts. And because that electricity is mostly doing work, not mostly giving off heat, it takes three to five times less energy to run. (Which is why even an EV plugged in to a coal-fired power plant is still far more efficient than an internal combustion engine.)

Better yet, consider an e-bike—my guess is that this will prove to be an even more important invention. A normal bicycle is already elegant technology, and an e-bike is essentially a bicycle without hills. It is almost unbelievably efficient—fully charging a 500-watt e-bike battery costs on average about eight cents. If that provides 30 miles of range, you'd be paying about a penny for every five miles of riding.

It's not, in other words, that an EV is a cleaner way to move your body and your possessions than an old-fashioned car, though it very much is. It's also better in every other way. If you like to go fast, an electric motor delivers maximum torque instantly. (Porsche's new electric Taycan can go zero to 60 in, um, 1.9 seconds.) If you like to avoid the mechanic, an EV is a great choice, as it rarely breaks. On a list of the top 10 most common repairs to an

American car, only one (replacing the fuel cap) applies to EVs, and even that's mostly aesthetic, since no one is going to siphon electrons out of your battery. (This new world takes awhile to sink in. When I was buying my Kia in 2019, the salesman, who clearly had not sold many electrics yet, offered to throw in six free oil changes to seal the deal. I looked at him for a long moment, and then he blushed, and gave me some floormats instead.) Even the big things that people worry about—that they'll have to replace the batteries, for instance—turn out to be wrong. In December 2024 *Wired* reported on two big new studies: "Rather than having a shorter lifespan than an internal combustion engine, EV batteries are lasting way longer than expected, surprising even the automakers themselves."

One of the EV's capabilities is almost magic. Instead of braking normally, when you take your foot off the accelerator in an EV, the energy from the rotating wheels feeds into a generator—essentially the motor is running in reverse. This regenerative braking means you won't wear out your brake pads any time soon; it also adds energy back into the battery, meaning you have more miles left in the battery at the bottom of a steep hill than at the top. The world's largest EV is a Swiss dump truck used to haul stone and lime to a cement factory. The empty truck climbs a 13 percent grade, is loaded with 65 tons of rock, and then drives back down, braking all the way. Each trip up the hill brings the battery down to 80 percent of capacity, and each trip back down refills it to 88 percent. It's about as close to perpetual motion as we're likely to get.

And it's not just trucks and cars and bikes. Crazily, work energy turns out be better than heat energy even at providing *heat*. Meet magic in a squat beige box—the electric heat pump is three to five times as efficient as the gas boiler that sits in most American basements. Essentially, as Stanford's Mark Jacobson explains, "It moves heat from one place to another instead of creating new heat"—that is, it takes the heat in the air outside and amplifies it to heat your room. (In the summer it runs in reverse to cool things down.) It's mostly pumping heat, not producing it.

Just as with EVs, there are urban legends surrounding heat pumps, chiefly that they don't work when it gets really cold outside. Since Finland has installed more of them per capita than anywhere else in the world this seems unlikely, and indeed the Department of Energy released a big new report in November 2024—their "residential cold climate heat pump challenge," the DOE reported, had revealed that the units "reliably provided heat with little assistance from auxiliary elements even during the coldest winter periods." The potential energy savings are enormous. Take Texas, for instance, where 60 percent of homes are warmed by old-fashioned electric resistance (think space heaters). If those homes, and all the other ones now heated by gas or oil, converted to heat pumps, the demand for electricity in Texas would fall, not rise, and the demand for natural gas would start to disappear.

It's such a huge and easy savings—you don't have to put in new pipes, you just install the inconspicuous blowers in key rooms, and the heat pump outside—that when Russia invaded Ukraine some of us responded by urging America to supply Europe with shiploads of the technology. (We have loads of spare capacity, because it's easy to convert an air conditioner factory to produce heat pumps.) We mounted a big campaign, and in the end President Biden agreed, diverting several hundred million dollars through the Defense Production Act. Like any new technology it takes work to convince people what's possible, but that work gets easier the more commonplace something becomes: In the UK, which has a goal of 600,000 new heat pumps per year to ease its dependence on gas, 400 families have joined the "Visit a

Heat Pump" program, inviting their neighbors in to see that the air is warm and the utility bills low. "Findings suggest that visits provide a final push and boost in confidence enabling visitors to accept a quote and progress to installation."

For households, the trinity of efficient appliances is an electric car or bike, a heat pump—and an induction cooktop. I'm not even going to try and explain how this latter appliance works (magnets!) but work it does, essentially turning the pan itself into the heat source. I'm the cook in our family, and I can tell you: once again, magic. You can boil water a third faster, even though you're using a third of the energy that a gas cooktop requires. And you don't have an open campfire in your kitchen. Cooking with gas increases the chances of respiratory illness in children by 20 percent—young lungs can't cope with the nitrogen dioxide and the particulates, and hence 13 percent of childhood asthma in the US can be traced to living with a gas stove. And to solve this problem you need about $60—that's what a perfectly good induction burner will run you at Amazon. Actually, $58.49 as of this writing.

In the US, something like 42 percent of the energy we use comes down to how we heat our air and water, cook our food, dry our clothes, and drive our cars. That is to say, almost half the emissions are the result of decisions we make around the proverbial kitchen table. A big part of Biden's Inflation Reduction Act (IRA) was designed to push those decisions toward the clean and efficient appliances I've been describing. But that's not easy—there are roughly 150 million dwellings in America alone, and we're talking about perhaps a billion machines that need to change. The IRA, at least before the Trump administration disfigured it, was designed to spend as much as a half trillion dollars over the next decade helping households replace those machines, *and all because the prize is so high.* Stanford's Jacobson calculates that if, by 2050, we'd convert the world to running on solar, wind, geothermal, and hydro power, we'd actually be using about 56 percent less total energy. A third of that savings would come from the switch to EVs, a quarter from heat pumps; hell, just mining, refining, and transporting fossil fuels requires 11 percent of all the energy humans currently use.

It's quite true that we're seeing big spikes in demand for power from data centers as artificial intelligence comes online, and in the places where those server farms concentrate, that will make for problems in this transition; this was the excuse for Trump declaring an "energy emergency" on his first day in office. But as we shall see, sun and wind make the most sense here, too, and the efficiencies from electrifying the whole system are so vast that even AI should be manageable. In fact, the newest frontier in electrification is not the home but the factory. A recent European study found that 95 percent of the industrial process heat in the continent could be replaced with electricity by 2035. Old-fashioned bricks may turn out to be a key part of this transition: It turns out that if you superheat bricks with solar electricity, they'll store that warmth for weeks, allowing it to be released as needed for making glass or chemicals, metal or cement. We're even learning to make steel with clean energy.

In 40 years of following the climate and energy debate, one of the funnier (and sadder) stories I ever read was written by Rebecca Leber, then a correspondent for *Mother Jones* magazine. In June 2020 she uncovered a massive campaign by the American Gas Association and the American Public Gas Association to pay Instagram influencers to gush about the gas flames on their kitchen stoves. There's a fetching picture of a young woman named Mrs. Jenna Martin in an orange skirt and a kind of black bandeau top (the better to show off her midriff tattoos) looking back somewhat alarmingly over her left shoulder as she cooks huevos rancheros. "Who's up for some breakfast for dinner?" she asks, adding, "Did you know natural gas provides better cooking results?" (Huevos rancheros, by the way, originated in 16th-century Mexico, 350 years before anyone was cooking with gas.) Leber found the pitch deck from one of the PR firms involved—it promised (with stock photos) to feature "snackable" content geared toward desirable millennial target audiences like "hispanic millennials," "design enthusiasts," "promising families," and "young city solos." (The stock photo for the latter, of course, was a guy with one of those Brooklyn beards.)

The gas industry has a long history of aggressive advertising—in the 1930s, trying to wean housewives off wood-burning cookstoves, it came up with the tagline, "Now you're cooking with gas"—but this new campaign had a very specific purpose. Some parts of the country were starting to pass bans on gas hookups in newly constructed homes and offices, trying to force them to go instead with induction cooktops and heat pumps. And that threat was best countered, they decided, with the allure of the blue flame. So Instagrammer "mayandtravel" wore her finest orange wool beanie as she cooked her favorite holiday dish, seafood noodle, over gas "because it helps cook food faster and gives me more control over the temperature." Right.

My point here is that, precisely because the conversion to things like heat pumps and EVs would cut energy use dramatically, the fossil fuel industry is going all out to make sure it doesn't happen. In fact, the entire point of the industry by this point in its history is to *make sure we keep burning something*. It's desperate, as we shall see, to slow down this switch by any means necessary—sometimes by bribing Instagram influencers, more often by buying political influence. It seizes on any event—the rise of AI, the fall of Ukraine—to insist that we prolong the life of its products.

Fifteen years ago, the thing the industry wanted to burn next was natural gas. Coal had acquired an unshakably bad reputation—it fouled the air over the cities where it was burned, and it produced huge amounts of carbon dioxide. The advent of a new technology called fracking (essentially exploding the underground geology so more gas will flow out) meant that there were suddenly large supplies of natural gas that could be easily substituted in coal-fired power plants; it became the bridge fuel, designed to carry us over till the day when renewable energy was ready to go. If you have any doubts about the power of this idea, read Barack Obama's State of the Union addresses—almost all of them include a paean to the fracking boom, which was both producing jobs and reducing carbon. Public research dollars, he said in 2012, "helped develop the technologies to extract all this natural gas out of shale rock—reminding us that government support is critical in helping businesses get new energy ideas off the ground." He wasn't alone—in 2009 that prophet Robert F. Kennedy Jr. hailed a fracking "revolution . . . over the past two years [that] has left America awash in natural gas and has made it possible to eliminate most of our dependence on deadly, destructive coal practically overnight."

The problem, it turned out, was chemistry. A researcher at Cornell University, Bob Howarth, started publishing a series of papers pointing out that, though burning natural gas produced half as much heat-trapping CO_2 as coal, methane (which *is* natural gas) itself trapped heat when it escaped unburned into the air; indeed, molecule for molecule, it trapped 80 times more heat than carbon dioxide at warming the planet. So, if your fracking and piping and burning let even a minuscule percentage of the gas leak out, it became a dirtier fuel than coal. And gas—well, gas is light. It's always trying to escape. The industry did its expensive best to tarnish Howarth and his findings, but study after study has shown that indeed large volumes of methane are pouring into the atmosphere. America has boasted mightily about how it's reduced its carbon emissions since the turn of the century, but since that mainly happened by replacing coal with gas in power plants and hence replacing carbon with methane in the air, it's pretty clear that we're damaging the climate just as badly as we ever did.

And we're not content with confining the damage to America. In 2015, while all the world's environmentalists were at the Paris climate talks, Congress lifted a long-standing ban on exporting oil and gas, and within a decade America was selling more gas to other nations than any other country. Late in the Biden years, environmentalists finally managed to make it an issue—Bob Howarth once again played a key role, with a paper showing that putting liquefied natural gas (LNG) on a ship led to so much leakage it was clearly worse than exporting coal. And so the Biden administration paused new permits for the giant terminals designed to feed gas to Asia. That in turn angered the oil and gas industry, which poured record amounts into Trump's campaign—on day one of his new administration he overturned Biden's pause, and by April 2025 it was clear that he was pressuring Asian countries into increasing LNG exports as a way to avoid tariffs.

Big Oil will do almost anything to stay in the burning business, because their reserves of oil and gas are currently worth tens of trillions of dollars. For instance, their advocates in Congress added tens of billions of dollars to the Inflation Reduction Act to pay for carbon capture schemes at gas-fired power plants. This defies economic sense. As I've already hinted and will describe in copious detail later in this volume, solar and wind power are not just better for the planet, they're also cheaper than energy from fossil fuels. Not "cheaper once you've figured in the costs of climate change." Just cheaper. So, if a solar farm makes electricity more cheaply than a gas-fired power plant, ask yourself how much more it will cost to refit that gas-fired power with a giant chemistry set designed to catch carbon dioxide as it pours from the smokestack and then pump it underground where it can safely be stored. The answer is, it will cost a lot more, so much more that even with massive government subsidies almost every project of this type has been abandoned.

Consider, for instance, Occidental Petroleum's Century project, one of the biggest examples to date of this carbon capture scheme. Occidental, remember, is the oil company whose CEO explained in 2024 that she planned to extract "every last molecule" of oil and gas beneath the earth's crust, a plan she said would be possible "if we can deal with the emissions." The Century plant, built in Pecos County, Texas, at extraordinary cost, went online in 2010; Bloomberg reported that Occidental quietly sold it at a quarter of its construction cost in 2022, after a career when it never managed to work at more than one-third of its capacity. Even when these plants actually operate, the carbon dioxide they "recover" is often pumped into local wells to "enhance oil recovery"—that is, to pressurize the ground so more crude can be extracted. Imagine if you'd spent the same money on solar farms and wind turbines that lowered the demand for gas; you'd "intercept" far more carbon, at far lower cost.

If simple economics dooms carbon capture schemes, physics and economics combine to make the next such step just as dubious. Billions are being spent—again, mostly of taxpayer money—to build out direct air capture, or DAC, projects, designed to scrub carbon directly from the atmosphere, instead of from smokestacks. (Occidental, having failed with Century, is now building a mammoth DAC plant called Stratos just a hundred miles away.) This task is actually even harder, since the flue gas in a power plant is 3 to 20 percent CO_2, as compared with about 0.04 percent in the ambient air. (As an MIT team explained in the fall of 2024, "The difference is akin to needing to find ten red marbles in a jar of 25,000 marbles of which 24,990 are blue versus needing to find about ten red marbles of which 90 are blue.") If the second job is too expensive, then you don't want to know about the first. And you don't want to know how much energy it will take: Using fossil-based electricity to run these machines, the MIT team said, would generate more carbon than it captured. If you used renewable electricity, Stanford's Jacobson points out, you'd be wasting the precise resource we need to power EVs and heat pumps—using a kilowatt of solar power to do something useful would remove seven to eight times as much carbon as using it to "capture" carbon. At the moment the largest of these direct air capture plants is in Iceland, where "dozens of huge fans suck air into bins that contain chemical pellets that absorb CO_2, which is then mixed with water and pumped more than a mile beneath the surface, where extreme pressure turns it into rock." But as *The New York Times* reporter David Gelles reported when he visited, the project—known as Mammoth—manages to capture one millionth of annual global emissions, and in May 2025 researchers reported it was not even capturing enough carbon to account for its own emissions.

The reason for these massive projects is less to capture carbon than it is to capture headlines—in the words of Australian energy analyst Ketan Joshi, they're designed to be "sprinkled on to fossil fuel projects as a sort of magical blessing . . . purely to fend off criticism and soothe investor nerves." It's all cover, like the huge ad campaign that Exxon ran for years explaining that it was planning to grow and harvest algae to produce carbon-free oil. When they launched the algae project in 2009 they promised it would be a "meaningful part" of the climate solution, and the *Times* described it as a "major strategic shift" for the world's biggest oil company. For the next decade and a half Exxon ran one commercial after another—you'd have been forgiven for thinking they were an algae company with some oil wells on the side. My favorite, perhaps because it recapitulated that tiny Ferris wheel that Bell Labs used to show the potential of the first solar cell, was an online ad that showed toy boats powered by algae circling a bowl. Someday, the voice-over intoned, algae could power "entire fleets of ships of tomorrow." But . . . it couldn't. A trial at a Welsh university showed that powering 10 percent of European transport would require growing ponds three times the size of Belgium. Exxon abandoned the effort (without any fanfare) in 2023; there was no world in which this actually worked, except the world it was designed for—the world of public perception. As one admiring PR industry newsletter put it, if the algae effort "burnishes the brand as it stares down rough headlines, or just softens the company's image in general—well, that's not small potatoes."

The effort to burn *something* will continue, always at great expense, and always under the cover that it's good for the environment. Ethanol advocates, for instance, are now trying to build giant pipelines across the Midwest, designed to carry the carbon dioxide from the refineries that turn cornfields into biofuel. That's despite the fact (which again we'll explore in more depth later) that covering a tiny fraction of those fields with solar fields would produce far more energy. Hydrogen advocates tout the idea that we can run trucks and buses on the fuel, which could theoretically be produced with renewable energy, though at the moment it's mostly made with natural gas. But economists keep cutting their estimates for how much hydrogen we'll need, as electricity from the sun and wind gets cheaper and cheaper. (The two worst-selling cars in the US both run on hydrogen—Americans bought exactly 89 Hyundai NEXOs in 2024, down from 241 the year before.) And it's not just the oil industry that wants to keep burning things; in Europe, utilities with old coal-fired power plants have retrofitted many of them to burn *wood*, arguing that as trees regrow they will suck up the carbon the power plants emit, making the technology carbon-neutral. This turns out to be perhaps the most delusional plan of all—the companies making the wood pellets are cutting down entire forests (mostly in the southeast US), which removes the most efficient machines we have for taking carbon out of the air; when the trees are burned, the carbon breaks the climate system *now*. That it may be removed 80 years hence when the forests recover is of no comfort at all.

Aside from sun and wind, there *are* other ways to produce electricity that don't require burning anything, and someday these may be useful. Hydro, obviously, though as I've explained, it's mostly built out now. Geothermal energy—which relies on the heat generated far underground by radioactive decay to turn turbines—may eventually get cheap enough to provide reliable backup power; in June 2024, a neighborhood in the Boston suburb of Framingham became the first in the United States heated and cooled by the warmth of its groundwater. The tides—a combination of solar and lunar energy—may eventually be harnessed. And of course there's nuclear power, which continues to soak up enormous public subsidy and continues to tantalize with the prospect of low-carbon energy. Nuclear power scares me less than it used to; I've spent the last decade arguing that we should keep existing power plants open. (It takes something going wrong to produce a nuclear disaster; whereas a gas-fired power plant operating precisely to spec destroys the earth.) But I think the prospect of atomic energy as a salvation is distractingly overhyped.

The latest burst of that ballyhoo came in October 2024 when various tech giants announced contracts with nuclear providers to provide power for data centers, but the headlines ("U.S. Plans to Revive Reactors as AI Powers Nuclear Renaissance") overstated the actual scale of the plans. Google made one of the biggest commitments, pledging to bring 500 megawatts of nuclear power to the grid by 2035. For context, the world currently installs 500 megawatts worth of solar panels every nine hours, and once you've connected them to batteries they provide "firm" power like nukes. It's not clear that data centers will even require power on the scale that AI apostles have predicted—the day after DeepSeek, the Chinese entry in this sweepstakes, reached the top of the Apple Store charts, utility stocks plummeted because it seemed as though training this new variant might require as little as a tenth of the electricity the early American entrants used. But even if we do need vast new server farms, solar offers the most likely method for powering them. A team of researchers from Silicon Valley stalwarts such as Stripe, Tesla, Terraform, and Anthropic produced a paper in December 2024 showing that the fastest, cheapest way for the Googles of the world (all of whom have pledged to slash their emissions) to power those new data centers was with co-located solar microgrids. "The tech is mature, the suitable parcels of land in the US Southwest are known, and this solution is likely faster than most, if not all, alternatives," the report concluded. Someone was paying attention; in December 2024 Google announced plans for $20 billion worth of solar farms located next to their new data centers.

That report on microgrids and data centers concluded with some strong language that I think applies to the planet as a whole: solar power, it said, was "likely the only clean solution that could also achieve the scale and speed required." Which is the point. The world has exactly one path available to make the rapid changes that the climate crisis requires, and that path is sun, wind, and batteries. They are available right now, in scale, at affordable price. And they can provide almost everything we need. In the future other technologies may supplement or even replace solar and wind power. It's possible that someday, for instance, we'll actually have fusion power, and it will be affordable. But that's not right now; right now money and attention spent on these glossy technologies is money and attention diverted from the task at hand. In April 2025, for instance, Spain announced it would slowly shut down its fleet of reactors because sun and wind generate "three to four times more power with the same amount of investment." Those who pine for nuclear power plants can console themselves with the reminder that the largest reactor in the solar system is sending us energy every time it peeks above the horizon. It's time to rejoice that we no longer need a bridge fuel—that the sun and the wind have built the bridge from the fossil fuel past to the clean energy future.

We're not impossibly far off the pace we need to hit. The International Energy Agency predicted in 2024 that the planet would double its renewable energy capacity by decade's end, installing more clean energy in those five years than in the last century combined. But the path that would take us to net zero emissions by 2050, the IEA said, requires not doubling but *tripling* that capacity by 2030. That won't be easy, especially given the steadfast resistance of the fossil fuel industry, now empowered by Trump's election. But hard is not the same as impossible, and Trump is not the first opponent; long before he was elected, the status quo was doing its best to prevent change. I remember standing on the tiled roof of a home in Phoenix in 2015, talking with Lyndon Rive, who is Elon Musk's cousin and, at that time, the CEO of SolarCity, then the fastest-growing solar installer in the country. We were looking out across the subdivision where his crew was working—we could see about a hundred homes, 11 of which boasted photovoltaic arrays. "It's like email in 1991," he said. "When I look out at this street, there's no reason every one of these houses can't have solar in ten years." Phoenix is after all in the Valley of the Sun. Its local college football team is the Sun Devils, and its professional basketball team is the Suns. It averages over 300 sunny days a year, making it one of the sunniest cities on the planet. Despite all that, the local utilities put up every conceivable roadblock for solar installation—they used ratepayer money, for instance, to campaign for the election of friendly candidates to the state's regulatory agencies. When those elections were over, the state's second-largest utility announced that it was going to charge $50 *a month* forever to anyone who wanted to put a solar panel on the roof. At the moment, about 12 percent of homes in Arizona have solar panels on the roof, compared with about 16 percent in cold and northern Vermont.

So it's not going to be easy, but it's what must happen. We have to bring burning to a halt—or we will burn. Consider Phoenix once more. In 2024 it recorded 113 straight days over 100 degrees, crushing heat records set . . . the year before. It was hot enough that the city's burn units were filled with people who had tripped and fallen on the sidewalk, where surface temperatures can routinely reach 180 degrees. As one of the city's doctors explained, for people who have been on the pavement for 10 to 20 minutes "the skin is completely destroyed." Here's how one survivor described his experience, which involved just a few seconds on his backyard patio. "When I was pushing on the rocks, they moved, but I didn't. I pushed until I couldn't stand anymore, and I looked at my hands and the skin had peeled off my palms like the skin of an onion. It looked like raw hamburger underneath, and I couldn't use my hands anymore."

2

Notes on a Different Kind of Burning

If global warming wasn't coming at us so fast, then it would be fine to let the force of economic gravity slowly make the changes we require, eventually eroding the primacy of gas and oil. We wouldn't need to force the pace, causing inconvenience. This is precisely the hope of most politicians, because fast-paced change is hard for any system to manage. I'll take as an example the widely read centrist-Democratic pundit Matt Yglesias, both because he's a talented writer and because he's struck this note so often. He's not a climate-change denier, but he criticized President Biden for calling climate change an "existential risk," and he insists over and over that "future industrial development is on track to be cleaner than past industrial development, even without any new policy changes or technological breakthroughs." Climate activists, he says, should pretty much shut up and let things take their course, because they scare normal people and create a political backlash. I'm not certain this political analysis is so astute (the Democrats won in 2020, when climate was a big issue, and lost in 2024 when no one talked about it at all), but I think Ygelsias's take is it's both widespread and plausible. Change is always easier, cheaper, and less traumatic when it comes more slowly—that's the most basic political reality.

And political reality is not to be taken lightly. But that's the problem: Climate is not an issue like most of the others that politicians deal with. Consider something that is a normal issue: health care. We should, like every other industrialized nation, provide it as a right. But if you can't find the votes for national health insurance, by all means don't scorn Obamacare; it helps, and the next generation will get its chance at instituting something more comprehensive. People may die and go bankrupt in the meantime, but it's not, in fact, "existential," at least for societies. It's a choice between different visions of how we should spend our money and what claims we have on each other.

Climate change, however, is not a choice between competing visions, at least not for the most part. It's only secondarily about political science. It's more about science science. So let's delve in.

It's worth remembering how recently we even came to understand the threat of CO_2—we only started measuring carbon dioxide in the atmosphere in 1958 (from a small hut on the side of Mauna Loa, the famed Hawaiian volcano). Biden and Trump were teenagers back then. We only started caring about the greenhouse effect in a public way in 1988, when James Hansen testified before Congress that global warming was underway. Since that time we've seen a steady ramp up in temperature, and a steady ramp up in the damage that elevated temperature causes. But the fear that drove international climate negotiations, and that spurred activists to get arrested, was mostly about the future. The dire events, from Hurricane Katrina in 2005 and onward, that marked this era were indeed dire, especially for the vulnerable populations (mostly poor people, and people of color) that suffered first and worst. But these events were still historically recognizable—they were amped up versions of things we'd seen before. They could, more or less, be absorbed by the communities and economies they smashed, though no one would argue that New Orleans is the same after Katrina as it was before. The real fear was what they *presaged*, especially since we just kept increasing the amount of methane and carbon dioxide in the atmosphere.

We are now in the "finding out" part of the exercise. For North Americans the realization should probably have started dawning in 2017, when Hurricane Harvey turned into the largest rainstorm (by far) in American history, dropping as much as five feet of water on parts of Texas, and in the process making it clear that warm air now held so much more water vapor that the old rules no longer applied. Certainly the penny should have dropped in June 2021 when, toward the end of the month, moisture from China mixed with heat from the American Southwest to create the most remarkable heat wave in recorded history (to that point) over the Pacific Northwest. On Sunday, June 27, Canada broke its all-time heat record, 113 degrees Fahrenheit, when the temperature reached nearly 116 degrees in Lytton, a community of around 250 residents on the Fraser River, in southern British Columbia. The next day, that record was broken, again in Lytton, when the temperature hit 118 degrees. On Tuesday, it was smashed again, when the temperature in the town soared to 121 degrees. On Wednesday, Lytton, now parched dry, burned to the ground in a wildfire; only a few buildings were left standing. Breaking a long-standing record is hard (Canada's old high-temperature record dated to 1937); surpassing it by 8 degrees is, in theory, essentially a statistical impossibility. It was hotter in Canada that day than on any day ever recorded in Florida, or in Europe, or in South America.

But those were preludes, and what's happened since has been even deeper, broader, and scarier. The last two years have seen that steady upward ramp turn into a vertiginous spike.

When 2023 dawned, I started getting calls from marine biologist friends. Pay attention to the temperature of the ocean, they said, because something unprecedented is happening. By late spring, ocean buoys off the Florida Keys were registering the highest sea surface temperatures ever recorded—101 degrees Fahrenheit, or more colloquially, the number you set a hot tub to.

And by June that heat was everywhere. Scientists have figured out how to take a daily temperature reading of the earth: There are enough thermometers on satellites and buoys and weather stations to produce an average for each 24 hours. That number is always highest around the summer solstice in the Northern Hemisphere, because there's more land to trap heat on that side of the equator. On July 3, 2023, scientists reported that day's temperature had broken the old record. In turn, July 4 and then July 6 set new all-time marks.

We were living through the hottest year ever recorded. Thermometers only go back a couple of centuries, of course, but scientists have found proxies in tree rings and glacial cores and ocean sediments that let them run the record back much further. They proclaimed quite confidently that those were the hottest days on our planet in at least 125,000 years. (To give some context, that's about the time that anthropologists think our ancestors started etching symbols on bone.) No civilization we'd recognize had ever lived on a planet this hot. And it's gotten hotter since—pretty much every month between June 2023 and the end of 2024 set some kind of global record, and 2024 was the hottest year ever by a considerable margin, smashing for the first time through the 1.5 degrees Celsius temperature barrier that the world had set as a red line just eight years earlier at the Paris climate accords.

This heat did extraordinary damage—heat waves across Asia in the springs of 2023 and 2024 produced stories of misery so frightening they hardly seemed real. A *Times* reporter, for instance, described a moment when one of the water tankers that were New Delhi's lifeline during the worst of the heat pulled up in a slum. "Men and women crowded around it, forcing the driver to stop. 'If you come close, I will slit your throat,' a broad-shouldered woman named Neetu shouted toward three women trying to snatch a water hose from her hand.

" 'Give me first,' cried a housewife, Geeta, who pushed Neetu to the ground.

" 'You have a grown-up family; my two children haven't had a bath for days,' another woman, Sarita, said while snatching the hose from Geeta. 'If you don't give it to me,' she continued, 'I will break this bucket on your head, then you won't be able to fill your bucket.' "

This strikes me as a working description of hell on earth. As do the now countless YouTube videos of extreme flooding, with cars washing down the streets of some mountain town. I live in Vermont, where we had the worst flooding in our history in the summer of 2023; my small town was cut off from the world for days when the Middlebury River ate huge bites from the two-lane road that connects us to the world. But no one died. When the deluge eased in late summer in New England, it picked up in North Africa, which had the biggest rainfalls in its history—in Libya, an enraged river smashed through two dams and then into a coastal city where in a matter of minutes it washed perhaps 10,000 people out to sea to drown. "Bodies are washing up on the shore every minute, on beaches as far as 150 kilometers away," the head of the local Red Crescent said. The manifest unfairness of this somehow makes it worse—the entire continent of Africa has produced barely 3 percent of the greenhouse gases warming the atmosphere. But it was eerie in a different (and maybe illuminating) way to watch the mansions of Pacific Palisades burn early in 2025. If you were imagining that you might get wealthy enough to spare yourself, think again.

The problem emerging over these crazily hot years is not just the daily trauma, it's that we seem to be doing deep damage to the planet's fundamental mechanisms. Back when that heat dome surged up the West Coast, for instance, scientists pointed out that one cause was a damaged jet stream. As the Arctic has steadily melted, the temperature differential between the poles and the equator has diminished, and that seems to stall the great currents of air in an exaggerated curve, plunging deeper and climbing higher in latitude. The melt of the Arctic Ocean and the Greenland ice sheet is also pouring fresh water into the North Atlantic, which seems to be disrupting the flow of the Gulf Stream and its associated ocean currents. That stream of warm water, a hundred times the volume of the Amazon, redistributes heat around the globe. It's what keeps Northern Europe warm; if it stopped, London would suddenly remember it's on the same latitude as Winnipeg, while tropical rain belts would likely shift dramatically. Scientists have long worried about this upheaval, but assumed it was a problem for the next century. Then, in a paper published in *Nature* in July 2023, a team from the University of Copenhagen used new data to estimate 2050 as the most likely date of such a widescale disruption—with the chance that it could come as early as 2025.

There are only so many really huge and vital systems on the planet, and now all of them seem to be in some sort of violent flux. In the Antarctic, recent heat waves have seen temperatures as much as 104 degrees Fahrenheit above normal—in late 2024, more than 450 researchers who study the region issued a declaration titled "Making Antarctica Cool Again" to point out that these ice sheets hold 150 feet worth of sea level rise, and to beg nations to "bend the carbon curve." Meanwhile, the earth's largest living structure—Australia's Great Barrier Reef, which stretches 2,300 kilometers along the Australian coast—has undergone mass bleaching events five times since 2016, with the highest ocean temperatures ever recorded. I've dived on that reef, into barren gray coral graveyards that just a few years prior were the teeming backdrop for David Attenborough's *Blue Planet*.

What else is on the list of crucial earth systems? The boreal forest of the north, stretching across Canada and Siberia, is the largest storehouse of carbon on earth. But the permafrost soils where that carbon is mostly stored are now melting ("drunken forests" of tilted trees are one witness), and in the last few summers huge swathes of those woodlands have caught on fire. In 2023 the blazes that choked American cities in orange smoke meant that Canada, with barely 40 million people, trailed only China, the US, and India as a carbon emitter. Around the world carbon emissions from forest fires have increased 60 percent since the turn of the century, an October 2024 study found. In the Amazon, where many of those fires used to be caused by loggers and ranchers, they were increasingly happening deep in pristine but drought-ridden "rain" forest.

In 2023 all of this added up to a temporary collapse of the earth's ability to soak up carbon. An international team of researchers found that the world's forests, plants, and soils had emitted as much carbon as they'd absorbed. "We're seeing cracks in the resilience of the Earth's systems," Johan Rockström, director of the Potsdam Institute for Climate Impact Research, said. "We're seeing massive cracks on land—terrestrial ecosystems are losing their carbon store and carbon uptake capacity, but the oceans are also showing signs of instability." Climate models had predicted this decline, but "happening rather slowly over the next hundred years or so," said Andrew Watson, who heads the University of Exeter's marine and atmospheric science group. "But this might happen a lot quicker. . . . The models are missing certain things." As 2024 drew to a close, polar scientists put out their annual "Arctic Report Card," and what they found was stunning—after millennia of northern lichen, moss, and grasses absorbing carbon, the arrow was now pointing the other way. "Between 2001 and 2020," the scientists reported, "wildfires and thawing permafrost caused the tundra to release more carbon dioxide than its plants removed from the air."

As researchers start to calculate the effects of trends like these, they're revising upward their prediction of how much the planet will warm. Those estimates had decreased some in recent years as nations began to tackle emissions, but more recently the United Nations Environment Programme revised its forecast upward to 2.6 degrees Celsius in 2022, 2.7 degrees in 2023, and 2.9 degrees in 2024. By April 2025, as they factored in Trump's radical moves, analysts at finance giant Morgan Stanley told clients they "now expect a 3°C world." Those may sound like small revisions, but research indicates that each tenth of a degree increase will move about a hundred million more people out of a useful climate niche and into "unprecedented heat exposure." By century's end, "roughly one-third of people worldwide could be outside the human climate niche"—if you think a few million immigrants and refugees have discombobulated the world's politics, prepare yourself for numbers a thousand times as large. As the *State of the Climate Report 2024*, issued in October, explained, "We find ourselves amid an abrupt climate upheaval, a dire situation never before encountered in the annals of human existence. We have now brought the planet into climatic conditions never witnessed by us or our prehistoric relatives within our genus, *Homo*."

So forgive me if I find the bland reassurance of writers like Yglesias unconvincing. "Nothing about climate change is world-ending or unavoidable," he writes. And yet worlds are already ending, entire and vast natural communities on which we all depend, and human communities we've constructed over millennia are now too hot to inhabit. This is an emergency, and it can't be solved by wishful thinking.

Oil companies with their algae ads and centrist politicians with their hope for a painless transition aren't the only wishful thinkers, however. There's also a strain of thought on the (vaguely) left traveling under the banner of "degrowth" that strikes me as equally unrealistic, and almost as likely to divert us from the work we must quickly do.

There's a "hard" version of this idea—that the world lacks the minerals to build out a future powered by sun and wind, and that even making the attempt will just result in a landscape-shattering last spasm of industrial extraction. This is unpersuasive for reasons I'll describe in more detail: As it turns out, weaning the planet off fossil fuels would mean far less mining and a sharp reduction in useless economic activity. Forty percent of the world's ship traffic, for instance, consists of moving coal and gas and oil back and forth across the ocean to be burned, a delivery job the sun accomplishes each morning as it moves across the heavens.

There's also a softer and smarter version of this idea, which argues that the pursuit of ever-larger GDP has turned into a trap for societies, and that we'd be extremely wise to reorient ourselves around measures like "gross national happiness" that attempt to balance lots of different goals. I find this persuasive for many reasons, including that climate change is the worst but not the only environmental crisis we find ourselves in. Twenty years ago I published a book, *Deep Economy*, that tried to consolidate this argument, pointing out that the greatest weakness of growth-obsessed societies is that they don't make us as happy as they claim to; that past a certain point (lower than one might guess) the relationship between "more" and "better" breaks down. I continue to find this logic compelling, but not as a reason to delay the project of converting the planet to sun and wind energy; in fact, as I will argue later, that conversion may be the most subversive possible step toward a more localized and modest world.

But there's a third version of this argument that worries me the most, and it's the idea that if we all simply lived more modestly we wouldn't need any of this technology, that we could avoid the need for building solar arrays and wind turbines and still head off the climate crisis. In my small Vermont county, residents are currently debating plans for a 300-acre solar farm. Among the common complaints is that we wouldn't need such an intrusion if only we lived more simply. "It used to be that environmentalists advocated for reduced consumption and living simply, but seems to have largely gone by the wayside," one resident of the county seat wrote to our vibrant local newspaper. "Perhaps they have resigned themselves to the endless tide of larger cars, larger houses, international vacations, and unlimited goods shipped to our doorstep." Instead of building a large solar farm, "we could put as much or more energy into figuring out how we might not need a large solar farm in the first place. Until we change how we live, tinkering with the fuel powering our consumption is unlikely to alter our planetary trajectory."

The reason this rendering of the argument worries me is that it's so sane. We manifestly don't need larger cars, and over a longer time frame (and aided, again, by the switch to local, smaller-scale sources of energy) I think we'll come to realize that. But I don't think there's any chance of it happening in the few years relevant to the outcome of the climate crisis: Remember, the world's climate scientists have told us we need to cut emissions in half by

2030, and people simply do not change their desires that fast. The very first environmental protest I ever organized was in 2001 in Boston—in the suburb of Lynn, actually, where a bunch of us marched down the strip of car dealerships, handing out flyers asking SUV shoppers to consider something smaller. The protest got lots of national attention, and indeed one of the banners carried by a minister—"What Would Jesus Drive?"—triggered a larger national campaign among evangelicals. It was all entirely logical—most sport utility vehicles never leave pavement—and it was also entirely unsuccessful. In the quarter-century since, cars have just gotten bigger and bigger; researchers now document the steady increase as "car bloat" or "autobesity," and these behemoths are spreading from America to around the world—in the UK, there are now 150 models too large for the standard British parking space.

Even if you could trim the excess of the Western world—and good luck with that since the current political temper of the Western world can be summed up as "You're not the boss of me"—there are, at this same moment, a huge number of people in the developing world becoming just a little like us. The World Data Lab projects 129 million of our fellow humans will "enter the global consumer class in 2025, with India (+47 million), China (+32 million) and Africa (+11 million) leading the way." I remind myself regularly of a conversation I had a couple of decades ago in a makeshift slum on what was then the third ring road around Beijing. I met an old man who had recently emigrated from the village where he'd lived his whole life to this warren of huts, and I asked him why he'd come. He gave me a big grin and said, "In my village there was no meat and no alcohol."

But again, even if you could convince everyone to make massive lifestyle changes, it wouldn't be enough to head off the gathering climate crisis. We know this because of a large-scale natural experiment called Covid-19. In March and April 2020, people across the world changed their lives far more profoundly than any environmentalist could ever have envisioned. Every plane was grounded; people hunkered down in their homes and stopped driving, stopped buying, stopped doing much of anything. And emissions did drop—but only by about 10 percent at the bottom of the pandemic. There's a great global machine undergirding our civilization, and at the moment it runs on hydrocarbons.

But it doesn't need to. Our task is to rip out the guts of that machine and replace them with sun and wind and batteries, and to do it while the machine is running. Over time, I think, that switch will make it easier to steer the machine in smarter directions, but that's not the main job at the moment. Emergency-room doctors don't waste a lot of time worrying about their patients' poor lifestyle choices—they do what they must to save their lives, perhaps with the hope that given a second chance their patients will choose more wisely.

We're very much in the ER—that's what all those statistics about the jet stream and the bleaching of the Great Barrier Reef mean. And I have to be honest: There's at least some chance we've waited too long already, that the climate die is cast. If the gnarliest predictions about how soon the Atlantic currents shut down or the Amazon turns to savanna are true, then all the solar panels we can imagine won't come quickly enough. I firmly believe that activists and engineers are the antibodies a feverish earth has summoned to save it, but I also know that sometimes fevers don't break and people still die.

Still, I'm going to end this dour chapter with a piece of good scientific news. Doomsayers have fixated on the idea that we're committed to a future of ever-escalating temperatures no matter what—that we're locked in to a pipeline of inevitable warming. But the latest science indicates that isn't true—that if we actually cut emissions of carbon dioxide and methane to zero, then warming would stop. As Imperial College London's Joeri Rogelj explained three years ago, "It is our best understanding that, if we bring down CO_2 to net zero, the warming will level off. The climate will stabilize within a decade or two. There will be very little to no additional warming. Our best estimate is zero." As Texas A&M climate scientist Andrew Dessler put it, "Some models show a few tenths of a degree of warming, and others a few tenths of a degree of cooling. However, our central estimate is that the global average temperature does not change once emissions stop." That's not euphoric news—we'd still inhabit an overheated earth. But no one goes to the ER expecting euphoria. What we hope for is a chance.

And that we've got, as we shall now see.

3

Life on the S Curve

If this book has a heart, we've now reached it. For this chapter I'm going to leave caveat and doubt behind and revel in the particular moment—conscious that we need to work hard to make sure the moment lasts, but convinced that sheer ebullience is one way to make sure that happens. Something staggering is occurring on our planet, something that wasn't foreseen. It is entirely possible that historians will look back on these few years and note them for the Trump renaissance, or the birth of AI. But if we're lucky, they'll be recalled as the moment we took a decisive turn toward the sun.

You'll remember that June and July 2023 shattered the old records for the hottest days ever measured on planet earth. Though the records are a bit imprecise, those also seem to be the months when humans passed a much more hopeful mark—for the first time we were putting up a gigawatt's worth (which is to say a coal or nuclear power plant's worth) of solar panels every day. By the Northern Hemisphere summer of 2024, when we once again crushed those temperature records, we were putting up a gigawatt of solar panels every 18 hours. We were suddenly on the steep part of the S curve, with wind and solar scaling faster than any energy technology before them. That gigawatt we were erecting every 18 hours? It had taken a full year to install it in 2004, and a week in 2016. Solar generation grew more in 2024 than coal-fired power has grown—total—since 2010. By 2026, according to the International Energy Agency, solar will generate more electricity than all the world's nuclear plants combined. By 2028 it will generate more than all the hydro dams. By 2030 it will have outstripped gas, and by 2032 coal. If we manage to stay on the IEA's net-zero-by-2050 curve, then solar energy will become the earth's largest source of all energy, not just electricity, by the 2040s. We're still at the very beginning of this boom. In 2024 solar panels covered an area just roughly half the size of Wales or of New Jersey. And yet they already provided 6 percent of the earth's electricity—three times as much electricity, as *The Economist* pointed out in a special issue on the "Dawn of the Solar Age," as the US used in 1954 when the solar cell was invented.

To say that this rapid growth was unexpected is an understatement. It would be more accurate to say that everyone who has tried to forecast how fast sun and wind power would grow have not just missed the mark but shot the arrow pretty much straight into the ground. *The Economist* again: "In 2009, when installed solar capacity worldwide was 23 gigawatts, the International Energy Agency predicted that in the 20 years to 2030 it would increase to 244 gigawatts." In fact, it hit that mark six years later, and now we're installing 244 gigawatts every six months. For most of the last decade the IEA's five-year forecasts missed by an average of 235 percent. The only group that came even remotely close to getting it right was not JPMorgan Chase or Dow Jones or Blackrock. It was Greenpeace, which estimated in 2009 that we'd hit 921 total gigawatts by 2030. We were 50 percent above that by 2023. It's possible that Jenny Chase—who has been tracking the economics of solar power for more than two decades for Bloomberg—knows more about the subject than anyone on earth:

"I simply can't believe where we are with solar," she told *The New York Times* in mid-2024. "If you'd told me nearly 20 years ago what would be the case now, 20 years later, I would have just said you were crazy. I would have laughed in your face. There is genuinely a revolution happening."

One reason we've missed some of that revolution is because so much of it is taking place in China, and we're used to thinking that anything important must happen in the West. By some measures, seven Chinese companies that I wager most Americans have never heard of—Tongwei, GCL Technology Holdings, Xinte Energy, Longi Green Energy Technology, Trina Solar, JA Solar Technology, and JinkoSolar—were by 2024 producing more energy than the Seven Sisters of the oil industry. "If Tongwei goes ahead with its plan for a 400,000-ton polysilicon plant in Inner Mongolia, nearly doubling its current output," *The Economist* calculated, "it might even overtake Exxon Mobil" (and Exxon is the company that did more to shape the world we know than any other). As longtime *Wall Street Journal* energy analyst David Fickling pointed out, even those numbers are an underestimate. "A solar panel sold by Longi in 2024 will be generating electricity for decades—most carry 25-year warranties. Oil and gas sold this year, however, will almost all be used up in a matter of months. If you look at the long-term flow of energy into the global economy that's crystallized with each solar cell produced, it's many times what's being provided by Big Oil." China, in 2020, set a goal of producing 1,200 gigawatts of clean power by 2030; in fact it hit that target in early 2024, six years ahead of schedule.

If you're wondering how this can be happening so fast, one answer is that this is fast tech. It takes, on average, way more than a decade to put up a new nuclear plant. The most recent American examples, Vogtle Units 3 and 4 in Georgia, took 15 years to build and came in at double their projected budget; by some estimates they are now producing the most expensive electricity on planet earth. A gas-fired power plant is a little easier—if you want a modern one with carbon-capture equipment, Stanford's Jacobson estimates it takes between six and 11 years from planning to operation. But energy from the sun is far more elegant. Even a decade ago, when I was clambering around rooftops with the SolarCity CEO Lyndon Rive, he was boasting about his crack teams that could do two rooftops in a day with time left to go surfing in the afternoon. The mammoth Horns Rev 3 wind farm, located in the North Sea off Denmark, took 22 months to build start to finish; it should last 30 years before it needs refurbishing, and when that day comes the refitting should take somewhere between three months and a year.

Even as we're building all this capacity, we're also—in the very same stretch of months—finally seeing EV and heat pump sales take off; by the end of 2024 more than half the cars being sold in China came with plugs. When we think "car" we think Detroit, but increasingly we should be thinking Changchun, and not just because they're producing *lots* of cars—they're also producing cutting-edge cars. In March 2025, China's leading carmaker announced it had cut the charging time for an EV to five minutes, barely longer than it takes to fill a tank at the pump. They're even producing *cool* cars. In December 2024, an article in the *Times* about how America was struggling to export automobiles included this truly striking paragraph buried midway through the story. It described BYD's new Yangwang U8 model: "The U8 can stay afloat for up to 30 minutes in a flood, according to the company, and its wheels can be set to roll so that the vehicle can rotate 360 degrees while remaining in the same spot." The Chinese brought some of their EVs to Las Vegas's giant Consumer Electronics Show in January 2025, and they drew long lines—the new Zeekr Mix minivan, for instance, gets not just 340 miles on a charge but also features "heated, ventilated and massaging seats that face each other and an interior that can transform into a lounge for card games with friends, a relaxing and comfortable fishing spot, or even a private yoga studio." The Xpeng AeroHT, which may or may not go into production, is a "six-wheeled plug-in hybrid meant to launch an electric mini-helicopter from its rear hatch." I mean, the Lincoln Navigator now comes with 13 cupholders, so there's that.

There's one final piece of the puzzle coming simultaneously to scale—the batteries needed to store energy when the sun goes down or the wind drops, so those e-bikes and heat pumps always have power. We—well, the Chinese—got skilled at building batteries for smartphones, and then used much the same technology for EVs; the price for lithium-ion storage has dropped 97 percent in the last 30 years. And now that expertise is producing big grid-scale batteries, so large that they can carry whole cities with their power. In 2025 the world will add 80 gigawatts of grid-scale storage, an eight-fold increase from 2021. America put up five gigawatts in the first half of 2024 alone; as the director of reliability analysis at the North American Electric Reliability Corporation put it, "Batteries can smooth out some of that variability from those times when the wind isn't blowing or the sun isn't shining. The Germans have a word for this sort of drought: *Dunkelflaute*. So, if you have four-hour battery storage, that can get you through a *dunkelflaute*."

Lithium-ion batteries keep getting better and better. In 2024 Chinese companies started producing EV models with 600,000-mile warranties, and if you eventually do need to replace your battery pack, by the end of 2024 that was less expensive than buying a new engine for a gas-fired car. And now the real excitement centers on even cheaper kinds of batteries that can store even more energy for every kind of purpose. Some are novel: In

Finland, for instance, a Scottish company is "converting a disused copper mine into a gravity battery." During the day, excess solar energy hoists a platform piled with tons of rock up a 530-meter shaft, and at night the platform lowers slowly back down, driving a turbine (as Euronews pointed out, the world is not short of abandoned mine shafts; indeed, they could store as much energy as the planet uses each day). You can use the afternoon's sun to pump water uphill too, or compress gas—that's what they're doing in Italy, where *The Economist* reports that carbon dioxide is being stored under pressure in giant domes—"When energy is needed the gas is expanded and passed through a turbine." In Minnesota, inventors backed by billions from Bill Gates and Jeff Bezos have pioneered huge, cheap batteries that use the energy produced as iron rusts. These are big enough to back up whole cities for four days at a time if there's an endless cloudy stretch; the first will be deployed in 2025 in California. Oh, and as we shall see, manufacturers are quickly learning to use sodium in place of lithium—sodium, as in the stuff that makes seawater salty. The first EVs with salt batteries rolled off a Chinese assembly line in February 2024, and by August Elon Musk was saying Tesla would follow suit.

The most important bottom line is that all of this remarkable technology is, finally, starting to undercut the endless growth in the use of fossil fuel. Ever since James Watt in the 18th century, the size of the economy has been closely tied to the amount of energy a society uses, and since until very recently "energy" and "fossil fuel" were synonymous. There's enormous momentum in this system. Because more and more of the world is finally growing its way out of poverty, energy use has just kept rising and rising—small energy drops in a Western world growing wary of climate change have been outpaced by the rapid rises of energy consumption in Asia. But now that's starting to change, even if, again, we can't quite see it for our underlying worldviews. Here, for instance, is my nomination for simultaneously the most hopeful and most stupid headline of 2024 (it comes from the normally astute editors at Bloomberg): "China's EV Boom Threatens to Push Gasoline Demand Off a Cliff." The writer is so used to the idea that we should use more gasoline that using less is a "threat"; of course, the news that the biggest economy on earth will use considerably less gasoline should actually be seen as an extraordinary gift. "The more rapid-than-expected uptake of EVs has shifted views among oil forecasters at energy majors, banks and academics in recent months. Unlike in the U.S. and Europe—where peaks in consumption were followed by long plateaus—the drop in demand in the world's top crude importer is expected to be more pronounced. Brokerage CITIC Futures Co. sees Chinese gasoline consumption dropping by 4 percent to 5 percent a year through 2030."

And it's not just China. Across Europe in 2024, emissions from the electricity sector fell sharply as renewables ramped up—fossil fuels provided just 28 percent of the total generating capacity, the lowest share ever. In Germany, emissions fell 11 percent, even though it was the first year since 1962 with no nuclear power plants in operation. But let's focus on Britain, where fossil fuel got its start. The UK has put up a lot of wind now, and that's one reason that in 2024 its carbon emissions fell below their level in 1879, a year that saw the start of the Anglo-Zulu War and the marriage of Queen Victoria's third son to Princess Louise Marguerite of Prussia. The British Library announced that it had turned to a solar thermal system to maintain the optimum temperature and humidity for their copies of Shakespeare's First Folio and the Magna Carta; King's College Chapel in Cambridge, where each year millions of viewers tune in for the Christmas service of lessons and carols, installed 438 panels on its 500-year-old roof. "We believe as Christians that the world is God's gift," the local bishop said. "We are called on as Christians to steward that gift." It all adds up: On the last day of September in 2024 Britain shuttered its only remaining coal-fired power station, at Ratcliffe-on-Soar in Nottinghamshire, with the blessings of the local union who said its workers had been offered job training. Some may end up working in what the plant's owner, a company called Uniper, described as a "low-carbon energy hub" to be built on the site. Closure in more ways than one.

When you're on a steep upward slope like this, novel things happen constantly. I've been randomly collecting stories that show the sheer exuberance of this breakout moment. Lots of them come from China, of course, where pretty much every day there's some announcement like "CHN Energy has connected the first solar units from its one-gigawatt offshore solar farm—the world's first and largest of its kind—to the grid," a project that will "power 2.67 million urban homes" from a field of 2,307 "solar platforms" five miles out to sea, each twice the size of a football field. (The project also hosts a fish farm.) But there's plenty of exuberance *everywhere*. In the US, 80 percent of new generating capacity in 2024 came from solar panels and batteries, and most of the rest from wind. In December, for instance, the State of Oregon approved its largest solar installation, and one of the largest in American history—10,000 acres, enough, according to its backers, to provide 800,000 households with their power (Oregon, for context, has about 1.6 million households and 61 million acres). Across the border in Washington state, work began in March 2025 on what will become the country's largest solar farm, 10,000 acres on the badly contaminated Hanford Nuclear Reservation. But the Pacific Northwest is blue. In purple North Carolina, the departing governor, Roy Cooper, closed out his term in January by boasting about the new clean energy jobs in his state: Toyota is building a $14 billion battery factory, Wolfspeed ("the world's only pure-play, vertically integrated silicon carbide company"), putting up a $5 billion plant; Natron is plunking down $1.4 billion for a factory to make batteries out of sodium. Or perhaps you want a red-state cocktail party fact: Did you know that the Western Hemisphere's largest solar panel factory was built over the past year in Dalton, Georgia, once the carpet capital of America, and the heart of Marjorie Taylor Greene's congressional district? (Inflation Reduction Act money is also building solar panel plants in Alabama, Florida, Texas, and Ohio.) "There are no words to describe how much better it is than a carpet factory," one worker at the Georgia plant told Politico. "My kids see me as a superhero. And I don't mean that to brag, it's about them saying that I can use the sun's energy for power."

Or consider Germany. Building codes were loosened in 2024, and almost immediately German apartment dwellers began to hang solar panels from their balconies. When the *Times* wrote about the phenomenon in midsummer there were half a million examples: "You don't need to drill or hammer anything," a 77-year-old retiree told a reporter as she installed three of the panels on the south-facing veranda of her Berlin flat. "You just hang them from the balcony like wet laundry in Italy." Apps "allow users to check how much electricity they are producing," the paper reported, and "for some that's become as addictive as social media or a video game." As one householder explained, "I am now completely hooked on how I can produce energy from the sun. It has become like taking a drug." Indeed, the trend just kept growing—by Christmas of 2024, according to *The Guardian*,

1.5 million Germans had hung up the panels, and some were producing a third of their household electricity demand; the country had just passed the 50 percent mark for generating its power from renewable sources, headed for 80 percent by 2030. Spaniards, Italians, and Poles were eagerly hanging out solar balconies too, and Belgium had just lifted a ban on the devices. Across Europe people are using solar panels for fencing, because they're now the same price as traditional wooden pickets.

In France, the government has mandated that every large parking lot in the country build solar canopies; in far-north Finland, the government closed its last coal-fired power plant in April 2025 and announced plans for seven new solar plants, including a big one on a bog currently used to produce that proto-fossil-fuel called peat. Norway's national football stadium, Euronews recently reported, now has the world's largest vertical solar roof, with upright two-faced panels that "can yield up to 20 percent more energy, making them valuable in climates with harsh and dark winters where maximizing energy production during shorter days is critical." Also, you don't need to shovel them after snowstorms.

Some of what's happening is silly (the EcoFlow Power Hat has a broad brim covered with solar panels that will charge your phone as you wander), and some is sublime—in the Ecuadorian Amazon, 12 indigenous communities now use a fleet of solar boats to get back and forth across their territory, silent electric motors replacing the gasoline that used to be flown in from Quito. ("The most common uses," CNN explained, are "transporting local children to and from school and providing wildlife tours for eco-tourists.") Some announcements involve the fastest-growing places on earth (in November 2024 the president of Indonesia said he'd close all the country's coal plants in 15 years) and some involve the smallest most remote communities (in May,

Argonne National Laboratory released a plan for replacing 95 percent of the diesel power at America's South Pole research station with sun and wind power). Some are visionary—Kamala Harris made an October campaign stop on the Gila reservation south of Phoenix to inaugurate a new project that shades irrigation canals with solar panels, thus producing power and reducing evaporation. (If you shaded all of California's canals with panels, by one estimate you'd save 63 billion gallons of water a year, enough for two million people.) Or consider Colorado, where the supply of clean energy is temporarily outstripping the capacity of the electric grid to carry it—while they're building new lines, the government is also contracting with a company called SunTrain to run a train filled with batteries from the solar fields to Denver every day. ("Trainsmission.") The Biden administration seemed to finally break the gridlock that was preventing offshore wind projects along the Eastern Seaboard; as a result New Bedford, Massachusetts, once the whale-oil capital of the planet, is reviving itself as a wind port (or was, until Trump's day-one moratorium on new wind). Meanwhile, some of the world's biggest cargo ships are adding rigid fixed "sails" to harness the wind for propulsion—the grain giant Cargill, for instance, now ships corn and soybeans to China in a 750-foot-long bulk carrier that uses 20 percent less diesel as a result.

Some of the plans announced amidst this exuberance are so grandiose they may never get built—in Australia, for instance, a local billionaire is investing in a project to use sun and wind power to generate green hydrogen for shipment to Asia. It would produce a lot of power—70 gigawatts, from 60 million solar panels and 3,000 wind turbines, as much as the size of the country's entire existing grid. "The proposal gives no starting date and is likely to take many years to work through the environmental processes," one skeptical journalist explained. But in the meantime there are already days when rooftop solar power alone is supplying more than 100 percent of power across South Australia; the country has 11 times as much solar capacity as it did a decade ago.

Giant companies are on board—by mid-2024, America's 10 biggest companies accounted for 17 gigawatts of solar power, roughly as much as all the solar power in the Sunshine State of Florida. Meta and Google and Amazon and Apple, obviously, but also Anheuser-Busch and Kaiser Permanente were on the list. But the action is as much on the micro as the macro end of the scale. Across the planet, for instance, 280 million e-bikes (and mopeds and rickshaws) were slashing oil demand four times as fast as electric cars. You might not even need to plug in—late in 2024 one company announced a solar scooter covered in enough panels to take it 30 miles (well, on a sunny day). "We were going for that Vespa quality of ride," says the inventor. "A lot of the inspiration came from the early advertising copy for Vespa. That was the vehicle that brought southern Europe out of World War II, and it was all about freedom and joy." Audrey Hepburn, on a Roman Holiday for the solar age.

After that sunny tour of the world, let's zero in one place: California, the fifth-largest economy in the world, slightly larger than India. And, in fact, let's zero in on one man, Stanford University professor Mark Jacobson. When Jacobson was growing up in Northern California in the 1970s, he showed a gift for science, and also for tennis. He traveled for high school tournaments to Los Angeles and San Diego, where, he told me recently, he was shocked by how dirty the air was: "You'd get scratchy eyes, your throat would start hurting. You couldn't see very far. I thought, Why should people live like this?" He eventually wound up at

Stanford as an undergraduate and then, in the mid-1990s, as a professor of civil and environmental engineering. By this time it was clear that visible air pollution was only part of the problem; the unseen gas produced by combustion—carbon dioxide—posed an even more comprehensive threat.

To get at both problems, Jacobson analyzed data to see if an early model wind turbine sold by General Electric could compete with coal. He worked out its capacity by calculating its efficiency at average wind speeds. A paper he wrote, published in the journal *Science* in 2001, showed that you "could get rid of 60 percent of coal in the US with a modest number of turbines." It was, he said, "the shortest paper I've ever written—three-quarters of a page in the journal—and it got the most feedback, almost all from haters." He ignored them; soon he had a graduate student mapping wind speeds around the world, and then he expanded his work to other sources of renewable energy. In 2009, he and Mark Delucchi, a research scientist at the University of California, published a paper suggesting that hydroelectric, wind, and solar energy could conceivably supply enough power to meet all the world's energy needs—a finding sharply at odds with the conventional wisdom at the time, which was that renewables were unreliable because the sun insists on setting each night and the wind can turn fickle. In 2015, Jacobson wrote another paper, published in the *Proceedings of the National Academy of Sciences*, showing that, on the contrary, wind and solar energy could keep the electric grid running just fine. That paper won a prestigious prize from the editors of the journal, but the acclaim didn't prevent more pushback—a team of 20 academics from around the country published a rebuttal, stating that "policy makers should treat with caution any visions of a rapid, reliable, and low-cost transition to entire energy systems that relies almost exclusively on wind, solar, and hydroelectric power" and accusing him of making "implausible and inadequately supported assumptions."

Which is why I went to visit Jacobson in midsummer 2024—I just wanted to hear him gloat. He met me in his driveway, in front of the light-filled two-story modernist house that he shares with his family at the end of a classic suburban cul-de-sac on the edge of the Stanford University campus. It's an energy-efficient showpiece; its solar panels produce more than enough energy to cover what he uses, though it is still tied to the grid. In the garage, there are two Teslas (including a 2009 Roadster with a license plate that reads "GHG Free") and a pair of the company's Powerwall batteries. The first place Jacobson showed me on the home tour was the mechanical room, where an air exchanger recovers 97 percent of the heat from the stale air that it pushes out of the house. Next up was the kitchen, where an induction cooktop cuts energy use by 60 percent compared with gas, even as it boils water twice as fast. As he made tea he showed me an app on his phone that monitors his usage of the power generated by solar panels on his roof every few seconds. "Yesterday, 13 percent of the generation from my rooftop went into the batteries in the garage," he said. "I used 8 percent of it at home, and I sold 79 percent to the grid."

Once upon a time renewable energy was the province of a few of us cranks who loved to keep track of how much we were saving (I started putting solar panels on my roof a quarter of a century ago, thrilled to see the meter run backward). But by 2024, in California, it had reached some kind of take-off point. Day after day, beginning in March, the state managed to produce from renewable energy more than 100 percent of the electricity it was using for long stretches of the day. I was following Jacobson's tweets documenting the surge—they were remarkably restrained, considering the abuse he'd taken for forecasting this precise breakthrough.

"Last year, we reached 100 percent a few times," he told me, as we sat in his living room. "But, this year, there's been 32 percent more solar output" as big new solar farms have come online, and "wind is up 11 percent." And demand for electricity from the grid has dropped 3 percent—mostly because so many people have put solar panels on their roofs, so they, like Jacobson, can supply much of their own power. "Tides have turned," Jacobson tweeted last week. "Fossil gas, coal, and nuclear are quickly becoming the 'alternative energy.'"

We Tesla-ed across campus to his office (he bikes when he's on his own) so he could show me his latest research. Essentially, his team tries to model what's needed for everywhere in the world to become California. So far they've produced plans to take 149 countries to 100 percent wind, water, and solar power by 2035. The latest countries added to his database, in the spring of 2024, were Madagascar, Rwanda, Uganda, and Eswatini (the former Swaziland). For each of them, Jacobson has a model that can forecast the weather every 30 seconds, for decades ahead, taking into account the predictions of a warming climate. If, on some June day in 2050, it's going to be 80 degrees in the mountains of Madagascar, and you want it to be 70 degrees inside a home, he can calculate the insulation value of the wall of an average residential building there and show how much energy it will take to cool things down. Then he can show several combinations of wind, water, and solar that will provide it. Very occasionally, he'll find a place with so little land that it can't produce the energy it needs on its own soil. (He limits the acreage to be used for solar and wind production to about 2 percent of a nation's territory.) "Singapore, Gibraltar, places like that," he says. "Then we go offshore." And, in the interest of grid stability, he tries to couple wind and solar in relatively equal amounts. "That's because in a heat wave, you have high pressure, and lots and lots of sun, but the wind tends to die," he says. "And then the low pressure comes in, and with it storms, which cuts the solar energy, but the pressure gradients mean strong winds." Hydro is a reliable source—essentially the biggest battery on the grid, because its power can be so easily stored for dispatch when needed—but when a drought causes its availability to drop, that almost certainly means that there's been a lot of sun. "Everywhere in the world, we can find ways to match demand for energy by supply and storage," he says.

And to do it affordably. Critics have pointed to California's high electric rates and blamed the sun and wind. In fact, "it's just the opposite," Jacobson said. California's prices have been driven up by wildfires, which are often sparked by utility wires, and natural gas disasters at San Bruno and Aliso Canyon. "If we didn't have renewables, our prices would be much higher," he said, handing me a sheaf of data to show that the other American states with high renewable penetration—mostly Midwestern wind giants such as Iowa and the Dakotas—have among the lowest electricity costs in the country.

If Jacobson had any complaints, they were that California was taking its foot off . . . I was going to say "the gas," but we need new tired clichés for the post-carbon age. The state's governor, Gavin Newsom, had recently ended incentives for rooftop solar, putting thousands of installers out of work, and—bowing to utilities and unions—making it harder to put panels even on the rooftops of schools and hospitals. And the academic in Jacobson hated the political compromises that meant the federal Inflation Reduction Act was funding carbon capture boondoggles as enthusiastically as heat pumps. "When the government is spending 40 percent of the money on things that aren't working, that's a problem," he said. "Most will fall from their own weight, but that doesn't prevent people from wasting a lot of time and money." Still, the view from California in 2024 was pretty remarkable, and continuing to improve. Thanks to the newly installed grid batteries, the state got through record heat waves in September without any blackouts (though the failure of a gas-fired power plant almost broke that record).

Sammy Roth, the climate correspondent for the state's biggest paper, the *Los Angeles Times*, ended 2024 by taking a chopper ride with Janisse Quiñones, the head of the city's Department of Water and Power, out past the edge of the vast city. (This was just before the fires struck.) "Downtown skyscrapers and Dodger Stadium parking lots gave way to lush foothill suburbs before our helicopter rose thousands of feet, cresting the San Gabriels' sharp peaks and trails. Rotor blades spinning, we swept north along Highway 14 into the high desert." From the air he could see the big solar farms, "a Tetris-like expanse of black wafers tilting toward the sun, surrounded by scrubby desert." When they landed and started driving, the latest addition—the Eland project that alone will supply 7 percent of the city's power—seemed to go on forever. But the real stars were 172 Tesla Megapack batteries, housed in "rows of non-descript white storage containers. They don't take up much space. But they soak up excess energy generated by the 700-megawatt solar project during the day and can disburse as much as 300 megawatts for four hours after sunset, sending the electricity coursing down a series of wires to the movie studios, port complexes, hidden data centers, and millions of homes that comprise L.A." Two nights before, with the project only half online, the battery pack had been sending enough juice back to the city to power 250,000 homes. The city is also getting power from wind turbines across the West, from pumped storage reservoirs north of the city, from geothermal plants in the Imperial Valley. It should be able to meet its goal of 100 percent clean energy by 2035, Quiñones said. And perhaps sooner.

"This is what every utility should be doing to fight the climate crisis: going big on solar and battery storage, while still investing in diversity," Roth wrote. And there are signs of that happening well beyond California. In fact, leadership in the renewable energy may be shifting from the City of Angels to the Lone Star State, which in 2024 was installing solar, wind, and batteries at a torrid pace—in the first eight months of the year it put up more photovoltaic panels than 39 states have *ever* installed. Texas is the world's eighth-largest economy—bigger than Russia. And it's also the spiritual heartland of fossil fuel. But it's easygoing regulatory environment, and its booming demand for energy means that the state will add twice as much clean energy as California and Arizona combined in the course of 2025.

If you want to know why, consider what happened on the evening of April 28, 2024. An early burst of summery heat meant that Texans were cranking the air conditioners, but as Julian Spector reported for Canary Media, more than a third of the state's gas-fired power plants were offline for maintenance. Just as things started to look shaky, however, "digitally controlled batteries rapidly injected two gigawatts of power" into the state's wires—the afternoon's sun and wind riding to the rescue after dark. By the end of 2024 there were evenings when batteries were supplying almost 10 percent of the state's power, batteries that hadn't even existed a year or two before. As one analyst explained, batteries follow sun and wind as—well, as night follows day. "It's a sequential reality of development right now," he said. "We expect in every market that deploys a lot of wind and solar, storage comes right behind it. You need shock absorbers, and storage is the shock absorber." By March 2025, Texas was setting new records for power from sun, wind, and batteries in the same week—and the solstice was still months away.

Texas is further from perfect than California; among other things, per capita energy consumption in Texas is more than twice as high. "It's because we have laws about appliances, we have strict building codes," said Jacobson of California. "It's all helped reduce demand." Whereas in bigger-is-better Texas, the only solution is more supply. And in Austin the state legislature, in thrall to Big Oil, has done what it can to throttle the revolution—but they run up against ever-more pushback from the consumers and corporations that like the cheap and reliable energy flooding on to the state's grid. In April 2025, legislators tried to set up a quota system to help the hydrocarbon industry—every megavolt of sun or wind or battery would have to be matched with a megawatt of gas-fired power. Rural residents flooded the state capital to campaign: "I'm here because I don't want people in Austin telling me and other rural landowners what we can do with our land," one man testified. "We have $30 million coming into our school district. It changed our community and it changed our family." Even the *Houston Chronicle*, hometown paper of the hydrocarbon industry, pointed out that sun and wind had saved the state's grid during an early season May heatwave, and asked the legislature to desist from "strangling wind and solar projects with the hope of reviving the natural gas market."

So the struggle's not over, even in the heart of Trump's America, simply because the logic of power from the sun is so overwhelming. On election day in November 2024, as the outcome became tragically clear, Jacobson tweeted out the news that once again California had managed to produce 100 percent of its electricity from renewables, even with the sun starting to dip in the November sky. For the year to date, he said, solar was up 26 percent from 2023, and wind power in California had increased

7 percent; electricity from batteries was up a staggering 97 percent in a single year. Better yet was what went down: Over the course of 2024, he reported on New Year's Day, *the state had used 25 percent less natural gas to generate electricity than it had the year before, a reduction that by the spring of 2025 had grown to 43 percent.* That's the single most hopeful statistic I've seen in 40 years of writing about our predicament. That's a number big enough to actually matter.

As good as this news is, it could get steadily better in the years ahead. As Jenny Chase, the superlative Bloomberg energy analyst, put it in the 2024 edition of her crucial book, *Solar Power Finance Without the Jargon*, "Human civilization is still in the 'shallow decarbonization' phase where solar and wind are nowhere near the fundamental limits of what they can supply."

You'll get a sense of what she means when you recall that the original Bell Labs solar cell was about 6 percent efficient at converting sunlight to electricity. At the moment, a new panel on the roof of your house might approach 20 percent efficiency. But if you read the trade papers for this industry—*PV Magazine*, say—you routinely come across headlines like "Novel bifacial PV [photovoltaic] cell offers 27% efficiency." That particular story, one of dozens for the week, explained that Indian researchers "have identified the optimal design with a Zr:In2O3 front transparent electrode, a CuSCN hole transport layer, and a NAN rear transparent electrode." A few weeks later you'll read that the European Solar Test Installation has confirmed claims from Chinese manufacturer Longi that their new "perovskite-silicon tandem cell" has hit the 34.6 percent efficiency mark. Perovskite is actually a word you may hear more of—it's a kind of calcium titanium oxide discovered in the Ural mountains in 1839 and named for an eminent Russian geologist. A perovskite solar cell has a similar crystal structure and they make highly efficient solar cells, but they degrade if they get wet. Researchers in at least

20 companies are busy solving that problem, in part by initially deploying them indoors—an Australian start-up says they're so efficient they can already power a set of headphones just on the light that comes from a desk lamp. They'll get better fast. As one academic predicted in the autumn of 2024, "I think we'll be pumping out rooftop solar by 2030," which would be nice since these cells may well hit 40 percent efficiency by then—meaning, in essence, they'd be twice as powerful as current panels.

Every part of the cycle is under constant improvement. *The Wall Street Journal*, for instance, recently reported on a new process, this one being pioneered by American manufacturers, that can dramatically cut the amount of silver in a solar cell by—well, by "coating the surface of plastic films with silver paste in precise patterns. That film then gets pressed onto preheated solar cells. The elevated temperatures make the pattern stick on the cell, so the film can then be peeled off." (This is happening at the plant in Marjorie Taylor Greene's district.) A week after that story, an Oxford team announced that they were in the early stages of dispensing with solar panels altogether, debuting a new coating 150 times thinner than a silicon wafer that could coat everything from rucksacks to cars, and was itself 27 percent efficient, though "the research team believes this could be extended up to 45 percent."

Perhaps there's a danger in describing these ongoing improvements—they might encourage us to wait for things to get even better. But, given the rising temperature, that would be exactly wrong. "We already have 95 percent of what we need," Jacobson told me. "Really, we can do everything except long-range aircraft right now," and that's barely more than 1 percent of emissions.

Instead, the challenge is less about research and development than it is about deployment—about constantly speeding up. Years ago I wrote a long essay about the last time we tried to go truly fast on a big technological project. In that case it was converting the American economy to war production in the months after Pearl Harbor. In 1941 the world's largest industrial plant under a single roof went up in six months near Ypsilanti, Michigan; Charles Lindbergh called it the "Grand Canyon of the mechanized world." Within months it was churning out a B-24 Liberator bomber every hour. Bombers! Huge, complicated machines, endlessly more intricate than solar panels or turbine blades—containing 1,225,000 parts, 313,237 rivets. Nearby, in Warren, Michigan, the Army built a tank factory faster than they could build the power plant to run it—so they simply towed a steam locomotive into one end of the building to provide steam heat and electricity. That one factory produced more tanks than the Germans built in the entire course of the war.

Mark Wilson, a historian at the University of North Carolina, has written the most comprehensive account of that period. It details how the federal government birthed a welter of new agencies with names like the War Production Board and the Defense Corporation; the latter, between 1940 and 1945, spent $9 billion on 2,300 projects in 46 states. "It was public capital that built most of this stuff, not Wall Street," he told me. It sounds, of course, a little like the Inflation Reduction Act, though in the latter case each public dollar seems to be triggering three or four times as much private money (there's a better long-term investment case for wind turbines than for tanks). Yes, Trump can screw things up (imagine if someone like him had been elected halfway through World War II). But the difference is that in this case America isn't the sole arsenal; in fact, the United States is playing catch-up to China, which a decade ago made the strategic decision to build green. The Chinese already have the factory capacity to build 1.1 terawatts of solar panels every year, which is actually slightly more than we need to hit the curve climate scientists are demanding, though in December 2024 the Chinese companies that own those factories announced an OPEC-like plan to rein in production in order to keep prices from plummeting further. It will be easier, for obvious geopolitical reasons, when more countries start to produce the key technologies. India, in 2024, was poised to become a net exporter of solar panels, for instance, and much of its production was headed for the US. But wherever they're built, the basic challenge is simple: deploy, baby, deploy.

My point is, we've done this before. And not just in wartime. We built the endless mammoth monolith that is the fossil fuel system: millions of miles of roads, billions of vehicles, gas stations on every corner connected to deep-sea oil fields halfway round the planet. All of that is much harder, technically, than what we have to do now—no German retiree can build an oil refinery on her balcony. We're not impossibly far off the goals we've set: At the global climate talks in Dubai in 2023, governments committed to reaching 11,000 gigawatts of renewable capacity by 2030. Right now we're on track to build about 8,000 gigawatts. "There is a gap," said one senior energy analyst at the International Energy Agency. "But the gap is bridgeable." In the rich countries, the IEA said, much faster progress on electrification is required; since EVs and heat pumps use so much less power, scaling them fast is critical. For emerging economies—think India—stronger efficiency standards for air conditioners and the like would help immensely. For the poorest countries, enough power for clean cookstoves is vital.

Our job is to flood the world as fast as possible with electrons from the sun and wind, confident that the very availability of clean, cheap power in bulk will drive the rest of the process. In the US alone, as *The Economist* pointed out, we have a terawatt of new solar capacity just waiting to be connected to the grid if regulators can get out of the way.

It would be unthinkable not to figure this out. The next sections of this book will get down to some of the details. In particular it will try to answer the objections that have arisen (Can we afford it? Are there enough minerals? Will it take too much land?) so that those critiques can't be used to slow this transition. But as we wade into those questions, I don't want to lose sight of the most basic truth: We are, quite suddenly, in a world-changing moment. We have a project, one with a deadline—indeed, we can echo JFK who said that "before the decade is out" we would land a man on the moon and return him to Earth.

Before our decade is out, we have to break the back of the fossil fuel system. We have to land the sun on the earth.

SECTION TWO

THERE'S NO REASON NOT TO DO THIS

4

Can We Afford It?

Of course we can afford it—the sheer fact that we're merrily building out terawatt after terawatt of solar and wind power is more or less proof that it's become affordable. But the idea that this is expensive alternative energy, as opposed to cheap fossil fuel, remains so deeply ingrained that we need to take a brief tour through the world of energy economics.

Speaking in Pennsylvania a few weeks before the 2024 election, Donald Trump repeated the standard gospel—wind, he said, "is the most expensive form of energy there is. You cannot get more expensive." It won't surprise you to learn he is wrong, indeed upside down. But environmentalists have helped build this misunderstanding. For most of the last four decades we've focused on making fossil fuel more expensive, with carbon taxes and the like—for the good reason that across those decades renewable energy indeed *was* too dear to compete head to head. But that's no longer true. All of a sudden this entirely necessary conversion is also the greatest bargain of all time, one that could upend our conventional economics of scarcity and replace it with something we don't yet fully understand.

Yes, of course, it takes money to build out this new system, and raising that capital is the great task of the next few years—but the quicker we do it the cheaper it will be, and when we're done we'll be repaid with a system so radically changed it might even begin to erode the cartoonish inequality that is our moment's other great threat.

First, though, let's reflect for a moment on what it would cost us to do nothing—that is, to continue lurching slowly through a haphazard transition off fossil fuels that happens too slowly to really arrest climate change. Everyone who thinks about it knows that the climate crisis will come with real costs. But in fact we don't much think about it, in part because the numbers that the models spit out are too large to make sense of. A few days before the election—right as Trump was spinning his tales about windmills—CBS News ran a valiant story making the effort to remind voters that climate change was already badly damaging the economy, and that the damage could grow far worse. "If the world fails to halt the rise in temperatures," CBS reported, "one study found that the impacts globally could cost $551 trillion by century's end, roughly 19 times the size of the U.S. economy." Indeed, that estimate for a century's worth of climate costs is *more money than currently exists on planet earth*. Yes, the economy will have grown larger by century's end, but a January 2025 report with the ominous title "Planetary Solvency," from the London-based Institute and Faculty of Actuaries, found that by 2070 the world could face a 50 percent loss in its GDP from climate shocks. On the current path, it said, the earth's systems could become so degraded that "humans could no longer receive enough of the critical services they relied on" to support our civilization. "You can't have an economy without a society," the report's lead author explained to *The Guardian*.

Against that scale of damage, our efforts at compensation are laughable. Forget, for the moment, trillions with a T. In May 2024 the World Resources Institute estimated that in the developing nations alone, the annual cost of the climate crisis will reach between $497 billion and $884 billion with a B before this decade is out; meanwhile, the world has set aside, so far, about $702 million with an M in a climate loss and damage fund to pay those countries. It's as if you'd set up a college savings fund for your newborn and plunked a nickel a week in the jar.

But the damage goes beyond those—highly speculative—numbers. The climate crisis is big enough to threaten the system those numbers attempt to describe, the thing we call "capitalism." Consider, for example, insurance, which is something we consider as little as humanly possible, since it's both boring and depressing. But it's also crucial: Insurance is the most basic underpinning of our economy, since if you can't hedge risk you won't do much. You can't get a mortgage from a bank, for instance, if you can't get insurance on the home you want to buy. And all of a sudden getting insurance is hard work. When, in the autumn of 2024, Hurricane Helene crashed into the Appalachians, it was carrying an ungodly load of water that it'd picked up over the record-hot waters of the Gulf—the radar readings near the summit of Mount Mitchell north of Asheville found that nearly four feet of rain had fallen. Not surprisingly, the cricks and hollers flooded. "We're seeing entire towns essentially flooded up to the first story," said Jon Schneyer, director of catastrophe response for the real estate risk analyst CoreLogic. "So we're talking total losses on properties in entire towns." In California, they'd just passed new laws in a desperate effort to keep insurance companies from fleeing the state when the Los Angeles fires broke out in January 2025; those blazes now seem likely to be the costliest disaster in American history, beating out Hurricane Katrina. In early February 2025, State Farm, the state's largest insurer, applied for a 22 percent emergency rate hike in order, it said, to send a message to "solvency regulators" and "ratings agencies" that the company had a "chance to begin rebuilding capital to sustain itself."

In December 2024, when the Senate Budget Committee was still in the hands of Rhode Island Democrat Sheldon Whitehouse, that august body released a mammoth report on the rising rates of "non-renewals" for home insurance. As wildfire and deluge have spread across the country, so have the number of insurance agents informing customers that their policies have been canceled: In more than 200 counties the rate of non-renewals have tripled in the last five years. The *Times* recently told the story of Lorri Williams, a recent widow living in a trailer home outside Silver City, Nevada, who got a cancellation notice in the fall from her Texas-based insurer. "REASON—UNSATISFACTORY RISK" the company wrote. "Your home is either located inside of or in close proximity of an area that is identified as having a high risk of wildfire." Her immediate problem, as the *Times* reported, was finding a replacement—her insurance broker could only offer the high-priced last-resort coverage offered by the state. Twelve percent of American homeowners had no insurance in 2024 at all, up from 5 percent in 2019, and as Bloomberg reported in December 2024, more and more of those who could find a policy were getting it from "non-allowed" companies—essentially unregulated firms that had been designed to cover "unique and relatively rare risks, like a fireworks factory or nuclear waste project." If you can get a new policy, the premiums often rise by a third or more—as the Senate Budget Committee noted drily, "This underscores that climate change has become a major cost-of-living issue for families across the country." Forget the price of eggs for a minute—insurance premiums are going up 40 percent faster than inflation.

But again, it's deeper than that. At some point—a point we seem to be nearing—the inability to buy insurance means that the value of homes begin to decline. The latest global estimate is that by 2050 climate change could wipe out almost 10 percent of the value of the planet's housing stock, or $25 trillion. For most Americans, one's home is the greatest source of one's wealth—therefore, as the Budget Committee reported, "any widespread decline in property values would thus present a systemic risk to the U.S. economy similar to what occurred during the 2007–2008 mortgage meltdown and ensuing global financial crisis." Indeed, "the difference from 2008 is that the financial system and asset values could and did recover. The physical risks of climate change make a similar recovery unlikely: a home too endangered to insure will only become more endangered." If you think other nations have figured this out, think again. In Communist China, Coco Liu reported for Bloomberg in December 2024, local governments have started to insure entire cities against flooding—but because the costs are so huge, the payouts are pointless. In Ningbo, where the average house is worth 23,000 yuan *per square meter*, a family stands to get a maximum payout of 10,000 yuan after a flood. *Per house*. (On average, "eligible households have received compensation under 660 yuan each.") Since warm air holds more water vapor than cold, today's floods are not the floods of yesteryear—there's simply not enough money to pay for the damage they cause.

And there's one more layer to this tearful onion. The actuarial table ranks high on the list of humankind's great inventions—by predicting the future, it allows one to build that future. Its only flaw is that it depends on the world behaving more or less as it has in the past. And that's now obviously a bad bargain, which means that at best there will be a lot more friction, in the form of ever-increasing premiums. But at a certain point it will all break down. I often remember an obscure report that the world's largest insurance firm, Swiss Re, wrote with experts from the Harvard School of Public Health way back in 2005. In the dry language customary for both actuaries and academics, the report concluded that before too much longer "in effect, parts of developed countries would experience developing nations conditions for prolonged periods as a result of natural catastrophes and increasing vulnerability due to the abbreviated return times of extreme events." That phrase—"abbreviated return times"—is the one that haunts me: it means blow after blow, delivered before you've recovered from the one before. It means that "development" gets harder and harder because you're just engaged in repair, and it means that eventually even the built-up wealth of the rich countries erodes. Where I live in Vermont, we had a great flood in 1939, and another in 2011 and another in 2023 and another in 2024, and so now, we're staggering.

Which is why it's so lovely to find an accelerating cascade headed in precisely the opposite direction, one that offers a way out. Let me introduce you to an American researcher named Doyne (pronounced dough-en) Farmer. Farmer grew up in New Mexico, a precocious physicist and mathematician. His first venture, formed while he was a graduate student at UC Santa Cruz, was called Eudaemonic Enterprises, after Aristotle's term for the condition of human flourishing. The goal was to beat roulette wheels. Farmer wore a shoe (now housed in a German museum) with a computer in its sole, and watched as a croupier tossed a ball into a wheel; noting the ball's initial position and velocity, he tapped his toe to send the information to the computer, which performed quick calculations, giving him a chance to make a considered bet in the few seconds the casino allowed. This achievement led him to building algorithms to beat the stock market—a statistical-arbitrage technique that underpinned an enterprise he cofounded called the Prediction Company, which was eventually sold to the Swiss banking giant UBS.

Happily, Farmer eventually turned his talents to something of greater social worth: developing a way to forecast rates of technological progress. The basis for this work was research published in 1936, when Theodore Wright, an executive at the Curtiss Aeroplane Company, had noted that every time the production of airplanes doubled, the cost of building them fell by 20 percent. Farmer and his colleagues were intrigued by this "learning curve" (and its semiconductor-era variant, Moore's Law), which show some technologies steadily falling in price and increasing in power; if you could figure out which technologies fit on the curve, and which didn't, you'd be able to forecast the future.

"It was about fifteen years ago," Farmer told me, in 2022. "I was at the Santa Fe Institute, and the head of the National Renewable Energy Lab came down. He said, 'You guys are complex-systems people. Help us think outside the box—what are we missing?' I had a Transylvanian postdoctoral fellow at the time, and he started putting together a database—he had high school kids working on it, kids from St. John's College in Santa Fe, anyone. And, as we looked at it, we saw this point about the improvement trends in renewable energy being persistent over time." That is to say, the steady fall in the price of energy from the sun since the first cell debuted in 1954 and the first wind turbine went up in the 1940s appeared to be an inherent feature of these technologies, something to count on.

But, as Farmer wrote in his 2024 book *Making Sense of Chaos*, this doesn't work for every technology, only those like airplanes or semiconductors where *technology* is the key ingredient. Coal isn't like that—for coal, a hunk of black rock is the key ingredient. You can get better at mining it, but on the other hand you start running out of the easy stuff to mine—those veins along the edge of the beach that once produced "sea coal"—and now you have to go farther down in the mine. Coal was cheap to start, but it hasn't gotten noticeably cheaper over time; ditto for oil and natural gas. The cost for generating solar power has by contrast plummeted, because it basically depends on people learning how to do stuff better. As the energy analyst Jenny Chase catalogued recently, the "relentless grind to lower costs" includes "better conductive pastes, less silicon wastage in slicing, thinner silicon wafers, better structural design of cells, and optimization of anti-reflective coatings and encapsulants." More specifically, she lauded developments like replacement of the Aluminum Back Surface Field solar cell with the Passivated Emitter Rear Contact, and the switch from slurry-based slicing of silicon to diamond-wire saws. You don't need to understand each of these things, only what they add up to: In 2014 *The Economist* held that "solar power is by far the most expensive way of reducing carbon emissions"; but a decade later, in 2024, *The Economist* put out a special issue devoted to solar energy that said, "An energy source that gets cheaper the more you use it marks a turning point in industrial history."

Sometime in those 10 years we passed some invisible line where producing energy pointing a sheet of glass at the sun became the cheapest way to produce power, and catching the breeze the second cheapest. That dive continues; indeed, the enraptured editors of *The Economist* now call it "the steepest drop in the price of one of the basic factors of production that the world has ever seen." It's a drop, they continued, that essentially faces no limit. "The resources needed to produce solar cells and plant them on solar farms are silicon-rich sand, sunny places, and human ingenuity, all three of which are abundant. Producing solar cells also takes energy, but solar power is fast making that abundant, too. The result is that, in contrast to earlier energy sources, solar power has routinely become cheaper and will continue to do so."

And once you start thinking this way, then your conception of energy economics changes dramatically. At the moment, we pay for fuel, over and over, our whole lives. The truck pulls into the driveway, the fat hose uncoils, the tank fills, the bill arrives. This is what made the Rockefellers rich—the simple fact that you have to write them a check every month for a new shipment. But solar and wind energy simply aren't like that. Once you've built the equipment to catch them—the solar panel, the wind turbine—*then the sun and the wind deliver the energy for free.* Yes, you need a battery to store some of that power, but the price of batteries has been plummeting right along the same curve as solar panels, and so even once you've added in energy storage, the cost is cheaper than fossil fuel. And falling, falling, falling, falling. If you understand that, then you understand the possible future. (And of course you understand why the fossil fuel energy companies will work so hard to slow that future.) But the bottom line is completely remarkable: Farmer's team at Oxford released a report showing that the rapid transition to renewable energy would, net, save the world 26 trillion dollars in energy costs in the coming decades. Because you don't have to pay for fuel.

The faster you do it the more you save, simply because you avoid the costs of all those visits to your driveway that truck otherwise makes. In September 2024 the International Renewable Energy Agency reckoned that "the cost of generating electricity from solar is less than half as expensive as the lowest-cost fossil alternative." In India, it costs less to put up a new solar farm than it does to buy the coal to fire an existing already-paid-for power plant. As the energy investor Rob Carlson put it recently, continuing to burn fossil fuel is a "self-imposed financial penalty" that will "ultimately degrade America's long-term global competitiveness. The same calculation applies to any nation, or any polity of any size, that choose to continue burning fossil fuels in any application in which electricity could instead be provided more competitively with renewables." This logic is so strong that even Saudi Arabia, the UAE, and Qatar are busily building vast fields of solar panels; in January 2025, OilPrice.com ("The No. 1 Source for Oil & Energy News") reported that by 2050 half the Middle East's power would come from solar PV, up from 2 percent in 2023. Even the Kentucky Coal Mining Museum has switched to solar panels for its power, in an effort to save 10 grand a year in electric bills.

Remember, not every energy source follows the learning curve that Farmer described. Not fossil fuels, and not, as it turns out, nuclear power. At least so far, he said, atomic energy is the "rare technology with a negative learning curve, becoming more expensive over time." Perhaps this will shift, if we start to mass-produce smaller reactors on an assembly line, but even then closing the gap with sun and wind will be hard work. As Farmer told me, "The only place on earth where you can find the cost of nuclear coming down is Korea, and even there the rate of decline is 1 percent a year. Compared to 10 percent for renewables, that's not enough to matter." The investment firm Lazard, which issues an annual report on the "levelized cost" of various energy technologies, found in 2024 that solar power was half the cost of coal and gas, and a third the price of nuclear. Farmer's team at Oxford found that "the nuclear scenario is by far the most expensive, with an average cost of $27 trillion" more than the fast transition to sun and wind. Even in France, with its long-term commitment to nuclear power, the cost of the most recent reactor ballooned to $14 billion; for that money, as *The Washington Post* pointed out in 2023, you could cover half the country's parking lots in solar panels and provide 10 times as much power. Which is to say, if we're ever going to get the 1950s dream of electricity "too cheap to meter," it will come from that nuclear reactor parked high in the sky.

It's hard to explain how different this is than the way we viewed energy economics until almost yesterday (and how most people, including those making crucial decisions, still do). The old idea was that the transition to renewable energy, though perhaps necessary to ward off the impossible costs of the climate crisis, would be ruinously expensive, and hence should be put off as long as humanly possible.

The wellspring of this belief was a paper by a Yale economist, William Nordhaus, published in *Science* in 1992 with the smug title "An Optimal Transition Path for Controlling Greenhouse Gases." It attempted to compare the costs and benefits of switching off fossil fuels, and concluded that we should make "modest investments," weighting "the costs of acting prematurely against those of acting too late" by employing the "tools of optimal economic growth." Under this framework, "stabilizing" the climate "would appear enormously expensive," costing the planet $6 trillion by 2100. Hence, we should use fossil fuels to get much richer in the decades ahead, at which point it would be easier to pay for a transition. The perfect path, Nordhaus later calculated, would see the temperature of the planet increase 3.4 degrees Celsius, 6.1 degrees Fahrenheit), which needless to say is far beyond what any climate scientist would recommend. If Los Angeles burns at an increase of 1.5 degrees Celsius, then we do not want to double down.

This framework was silly even back in those early days. For his 1992 calculations, Nordhaus drew on a paper he'd published a year earlier, in which he'd downplayed the economic risks of climate change because 87 percent of our GDP came from work in "carefully controlled environments," which is to say indoors. This was wrong on its face (you might write an insurance policy in an office, but the things it covered took place in the physical world), but it was also wrong because not all GDP is created equal. Writing code, for instance, is lucrative, but the people who write it still need to eat breakfast—breakfast is mandatory in a way that writing code isn't, and food to make breakfast depends on the outdoor temperature. I remember writing this precise critique all those decades ago, but I was not a Yale economist. Other economists eventually took issue with more technical parts of his argument—in particular, Nicholas Stern, an Oxford professor and one-time chief economist of the World Bank, who issued on behalf of the British government a 2006 report on climate economics best seen as a riposte to Nordhaus. It used a lower discount rate than Nordhaus, which meant he came up with much higher estimates of the damage that global warming would do, and hence he called for far swifter and more decisive action. Nordhaus attacked it as a "document that is political in nature," and reiterated his call to go slowly.

In the real world, Stern appeared to carry the day; his work was instrumental in helping build support for the Paris agreement, in 2015 still the high-water mark of climate commitment, with its call for reining in temperature increases by mid-century. But the world of economics—and of business—preferred Nordhaus's slow-walk version; he was awarded the Nobel Prize in 2018. And this kind of thinking stuck in important people's minds; in 2021, for instance, Bill Gates wrote a book describing the "green premiums" people would have to pay if they wanted clean power. By the time it filtered down to the brain of Donald Trump, he was describing renewable energy, as we've seen, as "the most expensive form of energy there is." And Donald Trump is, what do you know, president of the United States, partly on his promise to lower energy costs. What I'm saying is, we're anchored by an antiquated understanding that lingers even as it's proved wrong.

So let's try to construct a real-world alternative. I think often of a lunch I had in Glasgow during the global climate talks in 2021. I'd spent the morning in the city's main park, wandering with the novelist Kim Stanley Robinson, who had just published *Ministry for the Future*, the most important fictional account to date of the climate crisis. We'd been watching the start of the Fridays for Future youth march, led by environmental heroine Greta Thunberg. It was a sunny day, a carnival of mask and costume, music and chant; we followed around one group of fourth graders all in black with a sign that said "Ninjas for Climate Action."

And then I headed off to an upmarket pizza place along the main road in the university district, where over a very fine flatbread I sat and talked with Kingsmill Bond. In some ways you'd look at him and think: Nordhaus. Bond is a former investment professional, and he looks the part: lean, in a bespoke suit, with a good haircut. His daughter, he said, was that day sitting her exams for Cambridge, the university he'd attended before a career at Citi and Deutsche Bank that had taken him to Hong Kong and Moscow.

But he'd quit that world some years ago, taking a cut in pay that he's too modest to disclose. He'd worked first for the Carbon Tracker Initiative, in London, and then the Rocky Mountain Institute, based in Colorado, and as of 2025 the global energy think tank Ember. At the Glasgow pizzeria, amid the Parmesan and pepper flakes, he drew excitedly on a napkin, trying to explain to me what the numbers in that Oxford report of Doyne Farmer's really meant. Yes, we would have to build out the electric grid to carry all the new power, and install millions of EV chargers, and so on, down a long list—amounting to maybe a trillion dollars in extra capital expenditure a year over the next two or three decades. But, in return, Bond said, we get an almost incomprehensible economic gift: "We save about two trillion dollars a year on fossil fuel rents. Forever." Fossil fuel rent is what economists call the money that goes from consumers to those who control the hydrocarbon supply. Saudi Arabia can pull oil out of the ground for less than $10 a barrel and sell it at $50 or $75 dollars a barrel (or, during the emergency caused by Putin's war in Ukraine, more than $100); the difference is the rent they command. The mental world we've been inhabiting, he insisted, simply ignores those savings, which are particularly important for the people and places that can least afford to go on paying that guy to uncoil his fat hose in the driveway. Oh, and along with the money *you* save, it also comes with a chance at preserving an orderly civilization, which has to be worth something.

Earlier, I briefly quoted from the energy investor Rob Carlson, but I want to look a bit more deeply at his report, "The Sun Has Won," which came out in 2022, a year after those Glasgow meetings, because it makes the case in purely economic terms. "To the extent that local energy costs for manufacturing and services determine global competitiveness, the cost advantage brought by deploying low-cost solar will drive adoption in regions that wish to remain competitive," he began; in fact, even "continuing to operate the existing combustion fleet impairs economic competitiveness." (Indeed, a 2024 report found that it would be cheaper to close 99 percent of the coal-fired power in the US and replace them with sun and wind.) Carlson called this the "combustion penalty," and explained that though this is "frequently misunderstood by many journalists and investors, the vast majority of spending on renewables over the next several decades will simply be redirecting capital toward more efficient energy supply. That is, the large sums that will be invested in renewable power, and that are all too frequently portrayed as 'extra' spending will in fact be directed to renewables *instead* of fossil power." Instead of building new pipelines or replacing aging refineries, we'll build new solar farms.

There are lots of economic side advantages to this trend: Among other things, fossil fuel prices spike or crash whenever someone starts a war or even threatens one, and each time that happens it does its share of economic damage, while the sun comes up each morning no matter what else is happening. But the more fundamental difference between these two worlds is simply, as Carlson puts it, this: "Whereas the marginal cost of electricity production by photovoltaics and wind is approximately zero, to produce ongoing value from fossil-fueled electricity plants requires the constant incineration of assets, and hence the constant incineration of capital." Sun and wind are far more efficient than fossil fuels because they produce work not heat, and sun and wind are far cheaper than fossil fuels. Once upon a time coal and oil and gas were comparatively cheap. Carlson again: "Paying the energy efficiency penalties inherent in fossil fuel mining and combustion made economic sense because for centuries there were few competitive alternatives." So James Watt gave us a gift, albeit a complicated one. But now, "a dollar, yuan, or euro spent on fossil fuels delivers substantially less useful energy to the end user than the same amount spent on renewable energy." Given that, "the only uncertainty is how quickly, not whether, fossil fuels will be replaced by renewables."

As Kingsmill Bond kept charting these new realities on his olive-oil-smudged napkin, I was reminded of something that the liberal Canadian prime minister Justin Trudeau said, something that seemed at the time the height of a kind of cynical realism. He was speaking in 2017 at the annual gathering of the hydrocarbon industry in Houston; his topic was the vast tar sands in the province of Alberta, and I was paying attention because I was deeply involved in the fight against the Keystone Pipeline, which was designed to carry some of that extremely dirty crude south to Texas where it could be refined and exported. Trudeau, to a rare standing ovation from the oil crowd, declared, "No country would find 173 billion barrels of oil in the ground and just leave them there." In the energy economics of the old world, Trudeau's logic was brutally sensible. Never mind the toll the tar sands mining took on the environment; never mind the toll all that carbon would take on the temperature of the earth. (Burning that 173 billion barrels would mean that Canada, with 0.5 percent of the earth's population, would use up one-third of the world's remaining "carbon budget" outlined in the Paris accords.)

But as Bond pointed out, with lots of calculations, the brutal logic of extraction no longer was quite so obvious. Canada has fossil fuel reserves totaling 167 petawatt hours, which is a lot. (A petawatt is a quadrillion watts.) But, Bond said, Canada's potential renewable energy from wind and solar power alone is 71 petawatt hours *a year*. A reasonable question to ask Trudeau would be: What kind of country finds a windfall like that and simply leaves it in the sky?

That's especially true because converting to renewable energy also produces large numbers of valuable jobs, far more than the ones it displaces in the fossil fuel industry. Mark Jacobson's calculations, from 145 countries, predict 55 million new jobs in sun and wind, far outdistancing the 27 million lost in coal and oil and gas. And the better news is that the people currently holding the dirty jobs should be poised to get the clean ones; one study after another has found that, say, a coal miner has most of the skills necessary to install solar panels. I have fond memories of interviewing former oil field workers in North Dakota who were enrolled in a community college course to become wind technicians. "I enjoy big machinery, and it punched all those buttons," Jay Johnson told me. "They really are big, and, if you like machinery, then there you go." Most "of the job is general maintenance," he continued, "when you get up to the top of the tower and get into what we call the nacelle—it's basically a large gearbox and the generator and some control equipment. It weighs 80,000 to 100,000 pounds. So, there's a lot of changing oil filters, and lots of inspections, and, to everyone's chagrin, there's a lot of cleaning. You use a lot of Simple Green and a lot of paper towels." He added, "There's a lot of bolt torquing, too. You have to insure everything is nice and tight. Torquing and lubrication. And if it stops working, there's troubleshooting to figure out why it's not. That can be one of the more satisfying parts." It's also not boom or bust, like life in the oil patch; a wind turbine is there for decades, steadily and predictably churning away. You do have to deal with heights—a big part of the community college training had to do with ladders. But the new turbines usually come with some kind of lift, and employers have every incentive to make the work manageable. "It takes a lot of training to develop wind techs, and you can't afford to lose someone because they're long in the tooth. You've got to make things a little bit easier for them as they mature. And we're at a point now where people can retire as a wind tech."

If it all sounds too good to be true economically—well, that might be the only problem. Capitalism may be good at some things, but handling abundance may not be one of them; precisely because energy from the sun and wind is so plentiful and cheap, it can't make as much profit for investors as oil and gas, which are scarce and dear. A management consultant turned author named Brett Christophers made this case, persuasively, in a 2024 book titled *The Price Is Wrong*. As he explained to the American journalist David Roberts, "At the end of the day, we are relying not just on markets, we're relying principally on private sector actors to do this. And private sector actors are motivated predominantly, if not exclusively, by profit motivation. And it's expected profitability that drives investment decisions."

When the sun is really shining or the wind is really blowing, the spot price of electricity drops sharply—sometimes to zero. That's great in an enormous number of ways; it's obviously better for an economy, and the people who live in it, to pay very little for energy. But it's not so great for the investors trying to make money off energy. In Europe, in 2024, wholesale prices for electricity fell to zero (or below!) about 6 percent of the time; in California that number was even higher, about 20 percent of the time. There are ways around these fluctuations—batteries let you, in essence, rearrange time: noon becomes night. And you can incentivize customers to charge their cars and dry their clothes when power is most available, arbitraging time in the other direction. But this is a real problem. As Jenny Chase said in her 2024 report, "Low power prices may be great for consumers but they are very bad if you're trying to build more clean power plants." In December 2024, Bloomberg reported that "Sweden's wind power industry risks becoming a victim of its own success." I've biked across the countryside outside Stockholm, entranced by a spinning turbine around every ridge. But that buildout means "there's so much power around that electricity prices are increasingly dipping below zero, both for whole days and individual hours," and that in turn is "discouraging investors from backing new renewable developments as rock-bottom power prices offer little return," which in turn is threatening "Sweden's ambitious goal of reaching net zero emissions in 2045," five years ahead of the EU target.

All of this helps answer the question I get asked most often after talks: Why doesn't Big Oil just transform into Big Solar? It could, of course—Exxon has the cash flow to make huge investments in renewables. And from time to time some of these companies have made what seemed like serious efforts. I remember being at a press conference in 2000 when British Petroleum, one of the seven supermajors, announced that henceforth BP would stand for *Beyond* Petroleum, replacing its logo with a green and yellow sunburst it called Helios, for the Greek god of the sun. Its $200 million ad campaign featured slogans like "Beyond darkness, there is light," and "Beyond fear, there is courage." But it quietly dropped the new name in 2007, and soon was known mostly for the Deepwater Horizon spill in the Gulf of Mexico. It made another pass at greenness in 2020 at the height of Greta Thunberg's popularity, when it promised to cut its oil output by 40 percent by 2030 and rapidly grow its renewable portfolio instead—a project it abandoned in 2024. In March 2025 Bloomberg described the company's shift as "the tombstone for Big Oil's green pivot." BP is "now targeting new investments in the Middle East and Gulf of Mexico to boost its oil and gas output," in an effort to, as Reuters put it, "regain investor confidence." The CEO of Exxon concisely explained the underlying logic in March 2024—the company, he said, was going to keep investing in molecules ("and they happen to be carbon and hydrogen molecules") instead of in electrons because "we don't see the opportunity to *generate above-average returns* for our shareholders" with sun and wind. Exxon can control molecules—that's what "reserves" are, big pools of oil and gas. It can't control the sun, which just keeps pushing out power. And since Exxon can also control politics, at least in some places, it will stick to its current strategy, thank you. Eventually the cheapness of sun and wind will undercut that strategy, but everyone knows that businesses think in quarters now; and "eventually" lets the glaciers melt.

Christopher's insight also helps answer another interesting

question, which is why fully half of all the renewable investment underway around the world is happening in China. Clearly that country's leaders—who are somewhat less constrained by the logic of markets—have decided there are other reasons to make this transition. They know, for instance, how threatened their country is by climate change (the great manufacturing heartland of the Pearl River Delta is barely above sea level; Beijing struggles to find drinking water) and how tired their citizens had become of breathing foul air. They sense that global leadership may flow from taking the climate crisis seriously. They understand that economic dominance will come from building out these technologies.

And so, if we're to actually reap the potential benefits of this remarkable moment in time, we can't simply count on markets doing what's necessary by themselves. We need to figure out how to cooperate—preferably as a globe—to push hard. As we shall see, activists have plenty of work left to do.

5

But Can the *Poor* World Afford It?

One of the weirder small chapters in the climate saga came in 2010 when, as we've seen, the Obama administration began to push fracked gas as a way to reduce emissions. The nation's coal industry sensed the impending danger to their profits and suddenly found a heretofore unknown but definitely sincere and heartfelt commitment to the world's poor. Gregory Boyce, the CEO of Peabody, for instance, which at that time was the biggest coal miner on the planet, announced the "Peabody Plan to Eliminate Energy Poverty and Inequality" by—shocked face emoji—mining more coal. As Boyce explained to an appreciative audience in his keynote address at the World Energy Congress, the "greatest crisis we confront in the 21st century is not a future environmental crisis predicted by computer models, but a human crisis today that is fully within our power to solve. For too long," he explained, "too many have been focused on the wrong end game." For every zealot "who has voiced a 2050 greenhouse gas goal, we need 10 people and policy bodies working toward the goal of broad energy access." Only once we "have a growing, vibrant, global economy providing energy access and an improved human condition" for billions of the energy impoverished "can we accelerate progress on environmental issues such as a reduction in greenhouse gases." We can no longer, he stressed, "turn our heads from these brutal statistics. We must put people first." Mr. Boyce's industry colleague, Exxon CEO Rex Tillerson put it even more succinctly: "What good is it to save the planet if humanity suffers?"

As it happens, Peabody went into bankruptcy not long after (and court filings revealed that they'd been funding many of the world's biggest climate denial groups). Mr. Boyce went on to a career as the lead director of Marathon Oil. (He also serves on the board of Monsanto, that notable charitable foundation.) As for Mr. Tillerson, he accepted the Russian Order of Friendship award from Vladimir Putin himself before taking a job as Trump's Secretary of State, joining an administration that tried to cut hundreds of millions of dollars from humanitarian and peacekeeping missions.

Their basic contention about energy and poverty was both self-serving and wrong (coal, after all, had had 200 years to solve global inequality, a task at which it clearly failed). But the point has been echoed by less hypocritical sources—indeed, many of the leaders of developing nations have pointed out, sometimes scathingly, that the rich world has no business telling them to burn less fossil fuel. As the Senegalese president Macky Sall (eager to develop a big gas field), put it in a 2022 speech to the UN General Assembly, "It is legitimate, fair, and equitable that Africa, the continent that pollutes the least and lags furthest behind in the industrialization process should exploit its available resources to provide basic energy, improve the competitiveness of its economy and achieve universal access to electricity." That point seems unassailable; the US, with 3 percent of the world's population, has produced 25 percent of the greenhouse gases in the atmosphere. It has absolutely no moral standing to tell anyone else to stop using fossil fuels; instead, it owes a deep moral and practical debt to the rest of the world.

But of course the story doesn't end there. Because, first of all, if we don't get the climate crisis under control, it's not the United States that will suffer worst. And, more to the point, the developing world is not only suffering from climate change, it's suffering from having to pay for endless boatloads of coal and oil and gas. Senegal runs a trade deficit with the rest of the world—its biggest import is fuel. Oh, and a third of its citizens don't have electricity.

Eighty percent of human beings live in countries that are net importers of fossil fuel, which is one of those statistics that takes a minute to sink in. But as it does penetrate you start to truly understand the liberating possibilities of power from the sun. The sun shines everywhere, but most strongly and reliably toward the equator, in the direction where most poor people live. There are 940 million people on earth without access to electricity right now, and three billion who don't have enough. As the development expert Rajiv Shah wrote in the *Times* in the fall of 2024, "Access to electricity determines fundamental aspects of individual's lives, like whether they are healthy, or have a job." But "improved solar panels, batteries, and other breakthroughs now make it far easier to provide reliable, clean electrification to everyone." If you want to end poverty, he said, focus on this one thing and nothing else. Do it right, the editors of *The Economist* said in the summer of 2024, and "much of the world—including Africa, where 600 million people still cannot light their homes—will begin to feel energy-rich. That feeling will be a new and transformational one for humankind."

So let me tell you a story—not a story that shows how to solve all our problems, but a story that shows some real possibilities. It comes from Pakistan, which is arguably the country on this planet hardest hit by the climate crisis. Its towns and cities have recorded some of the highest temperatures ever seen on this earth—in May 2024, for instance, in the Sindh province town of Mohenjo Daro, where archaeologists say humans have lived since at least 2500 BC, the mercury crossed 128 degrees Fahrenheit; in the city of Jacobabad, workers had to knock off by midday, cutting their wages in half. "This isn't heat," one brickmaker told reporters. "It's a punishment, maybe from God." But the heat's the least of it; twice in the last 15 years Pakistan has suffered floods unequal to anything since Noah. In 2022, that same Sindh province (home to 55 million people) saw August rains 784 percent greater than average—the kind of relentless rain that only a globally warmed world can produce. Mud huts eventually simply melted in the deluge; 897,000 homes were wrecked. The country's climate minister reported that a third of the nation was flooded, and 33 million people had their lives turned upside down. All in a nation that has produced less than 1 percent of the planet's greenhouse gases.

But climate change is not really what drives the story I'm eager to tell. Beginning in 2023, energy analysts started noticing something bizarre: Demand for electricity on the national grid had begun to fall, and substantially—Pakistanis seemed to be using a tenth less electricity. But that doesn't happen—demand for energy, except in very deep recessions or crises like the Covid lockdowns, always goes up; that's been the basic fact of life since James Watt. So what was happening? The answer began to emerge over 2024, as analysts started studying images from Google Earth: big expanses of solar panels were sprouting on top of apartment buildings, factories, shopping malls. These weren't government-sponsored installations that came with press releases and ribbon-cuttings. These were just people deciding that they could produce their own power more cheaply and more reliably than the national grid. Chinese factories had produced so many solar panels that there was a glut on the market, and it was relatively easy for that glut to show up next door in Pakistan. Panel prices dropped 60 percent in a single year, and all of a sudden traders who'd been importing rice and cement started pushing solar panels. And they had no problem finding buyers. "Now we never face the problem of power cuts, and our power bill is almost nil," one 34-year-old young professional told Nikkei Asia. "Now I can get peace of mind and focus on my business with uninterrupted access to electricity all day long and don't have to worry about load-shedding," a medical equipment dealer in Lahore explained.

Some of the switch is driven by outside demand—lots of large Pakistani textile firms are switching to solar because Western apparel brands demand "sustainability," and the biggest supplier of soccer balls for Adidas said he was feeling pressure from the company to "go green." "Every bit of space I have, even if it's a few feet, I want it covered in solar panels," the factory owner, Khawaja Masood Akhtar, told the *Financial Times*. "It's the only way we can beat our competitors. Allah has given us this gift." But mostly it's just the obvious economic advantages of solar that are winning over customers, many of them poor. As one Lahore-area corn farmer, Mohammad Murtaza, told the Indian financial paper *Business Standard*, "I have never seen such a big change in farming. Ninety-five percent of farmland has switched to solar in this area," he said, pointing to his photovoltaic array towering over piles of harvested corn cobs. Many farmers can't afford the metal mounting brackets, which are more expensive than the panels—they just lay the panels on the soil, cells to the sun. If you've traveled through rural Asia, you know the sound of diesel generators pumping the millions of deep tube wells that were a chief driver of the agricultural Green Revolution of the

1960s and '70s. Now it's solar electricity that is pushing up the water. Diesel sales in Pakistan apparently fell 30 percent in 2024, largely because the panels were providing the energy. If you're a farmer, that's kind of a miracle; one of your biggest costs is simply gone. And if you're worried that a power cut is going to shut down your fan in the 128-degree heat, it's a kind of miracle, too. As a Pakistani solar entrepreneur told American journalist David Roberts in February 2025, "A 3-kilowatt inverter with, you know, maybe four or five panels" is now routinely included in a bride's dowry.

The speed at which this miracle is taking place is almost unnerving. In the first six months of 2024, Pakistanis—again, working on their own, without government guidance—installed the equivalent of 30 percent of the nation's electric grid.

That means, of course, that the story comes with complications. Some of the panels may be lower quality, or counterfeit. And the national grid is suddenly in trouble—there's talk of a "doom loop" for the utilities, which (ironically) built big and expensive power plants with Chinese loans. This may, at least in the short run, harm the very poorest Pakistanis, who depend on the national grid for the small amount of power they use and can't afford even the cheap panels. But all of that is overshadowed by the remarkable shift underway—as energy analyst Azeem Azhar explains, "The switch is profoundly pro-local and pro-entrepreneur; they no longer have to rely on a government for your energy supply—they can supply it themselves." Not only are Pakistan's power-sector CO_2 emissions now falling as the country uses less coal, oil, and natural gas to produce power, but the possibilities for average people are opening up in unexpected ways. It's not so different from the moment 20 years ago when mobile phones suddenly made landlines irrelevant in the developing world. "Self-sufficiency could grow from the bottom up," Azhar writes. "Corrupt officials who parlay their power over national infrastructure and policy into bribes have one less level. As demand for oil and gas drops, energy traders have fewer opportunities to make bank. That loss signals a broader gain. The silent energy revolution in Pakistan isn't just about keeping the lights on; it's a fundamental restructuring of power—both electrical and political."

What Azhar calls "energy self-sufficiency," or what American politicians have long called "energy independence," has been a great boon for a few countries that happened to have deposits of coal, gas, and oil. But now pretty much every country can join that club, and their citizens can reap the benefits. If the temperature reaches 128 degrees—well, at least that's a good sign you're well-endowed with sunshine.

If there's any place that the concept will be put most fully to the test, it's Africa.

It's easy to imagine that the developing world is only now getting a taste of clean energy, but that's not actually true; in fact, I'd wager more Africans than Americans have practical experience with solar power. I missed the first wave on the continent—the only glimpses I've seen are in a small museum of sorts at the University of California's Humboldt campus. Its curator was a professor named Arne Jacobsen, who had been in Kenya in the 1990s and watched the very earliest photovoltaic power come to cities and towns. Much of the technology had, says Jacobsen, "big troubles. Chinese panels, panels from the UK, all this low-quality junk coming in. Later, LED lights that failed in hours or days instead of lasting thousands of hours, as they should. People's first experiences were often really bad."

I was on hand, in both West and East Africa, for the next

stage of the solar revolution, in the mid-2010s. Money, especially from Silicon Valley start-ups, was pouring in, and start-up companies were setting up fleets of salesmen on motorcycles capable of handling the continent's backroads. I watched as they sat on the ground outside huts in remote villages, pitching people on solar panels as those potential customers pounded clothes in tubs. If people bought a tiny system—one panel, an inverter, a few outlets—they paid a monthly fee (through their cellphone). It was often less than what they were paying for the kerosene in a sputtering and smoky lamp, and as long as they kept up their payments they could summon another fleet of motorbikes carrying technicians if something went wrong. Watching these installations was uniformly moving—what do you say when a man is tearfully explaining that now his daughter can watch for snakes outside the door to the family home? It was also of course slightly unnerving; l started encountering whole villages where everyone was watching the same Bollywood soap opera, or had somehow, with the advent of Premier League broadcasts, all become Arsenal fans. "In the old time, you had to go outside and talk," one old man told me enthusiastically. "Now my neighbor has his TV, I have my TV, and we stay inside."

One striking thing was the almost immediate demand for more power. I was in a one-room house in rural Côte d'Ivoire where a man was showing off his lightbulb and his radio—but what he really wanted, he said, was a fan. Crime was bad enough that he closed his window at night, and then it really sweltered in the closed room where he slept with his wife and daughter. "I work outside all day," he said (this was a cacao region). "It would be so relaxing to be cool at night." More power means more panels, and as those Chinese factories have started churning out ever-more-affordable versions, something like the Pakistani story is also playing out across Africa. In the middle of 2024 I had a long and interesting conversation with Joel Nana, a Cape Town–based energy analyst, who was struggling, like the Pakistan-watchers six months earlier, to understand new data. "In Namibia we've uncovered that they have about 70 megawatts of distributed generation, mostly rooftop solar PV—that's the equivalent of about 11 percent of the country's grid. In Eswatini, which is a very small country, it's about 15 percent." In South Africa, the continent's economic colossus, that number is about 9 percent. "You won't see these numbers anywhere," he said. "They're not reported in national plans, no one knows about them. It's only when you speak to the utilities. And in fact the numbers could be much higher, because the smallest systems aren't reporting to anyone, not even the utilities."

Here again the switch is being driven by the desire for reliable and affordable power. In April 2024, for instance, Nigeria's electric grid had its fifth collapse of the year, a few weeks after the utility had raised electricity prices by as much as 230 percent. Nigerians survive because they have backup diesel generators; indeed, those "backup" generators can supply far more power than the national grid. But it's expensive to keep pouring diesel in the tank (and God they're noisy), and so as PV prices dropped, "solar has become a no-brainer for most businesses if not all; the prices just make sense," says Nana. "In a lot of places, it's all the malls, all the mills—any business that has enough roof space." Africa, like Pakistan, had well-established trade networks with China, and so the panels have come flooding in. "You have some utilities, like in Mozambique or Madagascar, that see it as a threat, and are trying to claw it down. But the realization is, this is happening anyway, whether you like it or not. If you fight people, they'll just go clandestine and install it without letting you know." Instead, Nana and his colleagues are concentrating on helping governments make the most of the boom—to come up with enough structure to turn the patchwork into a system, but not so much that it dries up the entrepreneurial vitality driving the change. "We can think of a way of empowering communities through the local ownership of these assets," Nana said. "So people like me can be empowered to start having transactions—I could be allowed to wheel my electrons to someone in a less-affluent suburb." In cities, local governments could allow urban residents to install as much solar PV as they can, and that could "support the government's efforts in providing more energy access and generation for people in more remote areas."

The last time I was in Africa, the most visionary entrepreneur I met was an African American woman, Nicole Poindexter. While the Silicon Valley money was funding those motorcycle fleets of salesman pitching panels for people's roofs, she was experimenting with what has since become the most important vehicle for extending power—the mini-grid. It's really just a solar farm on the edge of a village, with enough rudimentary wiring to connect the houses, but it offers enough scale to lower prices and to provide more power. The first ones she showed me, in northern Ghana, were revelatory: towns with clinics where the doctor could suddenly refrigerate vaccines and deliver babies at night without a flashlight clenched in her teeth, towns where there was enough energy to run a boombox and big speakers for nighttime parties. "Our relatives from the city used to not come here to visit," one local chief told me. "Now they do."

Poindexter was ahead of the curve, and she's stayed there—now she mostly works in Sierra Leone, and with communities 10 or 20 times bigger, less villages than towns and small cities. She's part of an ever-expanding network—as Tombo Banda, a Malawian who runs a Mini-Grid Innovation Lab, put it in a recent interview with David Roberts, "We are really building this future grid now, which is more of an . . . interwoven mesh of 'main grid,' mini-grid, solar home systems, electric cars, charging networks for electric buses." At the moment, she says, it's hard to attract private investment because the people who need the mini-grids are poor and can't pay much; as they get power, they will grow wealthier, and they will in turn use more power. "We're going into a community, we're building a mini-grid, and we're going to figure out ways of increasing the revenues we get from each consumer in that community. And how we do this is by providing them with electrically powered income-generating machinery that will [allow] them, you know, if they are farmers to do agro-processing on our electrically powered machinery." This all could work because—just as with EVs or heat pumps in the West—the electric appliances run more cheaply than the diesel-powered ones they replace; they produce work, not heat.

But there's an obvious chicken-and-egg problem here. Africa has about 3,000 of these mini-grids now, and it needs about 160,000 to really supply the continent, which Banda says would take about $90 billion. That's half a million per mini-grid, which is not much in the scheme of things—a mini-grid is basically a couple of container-loads of gear. But we're talking about communities without capital—we're talking about a continent where the wealth has largely been extracted and ended up elsewhere. The capital that's necessary for the one-time investment to liberate the sun ended up, above all else, in pension funds in the West—the retirement funds for public employees and teachers in California, New York, Texas, and Florida are all in the world's top 20, each with hundreds of billions of dollars; heck, Minnesota's public employees have a retirement fund of more than $100 billion, which would be enough to get all those African mini-grids up and running. But if you spent your life corralling fourth graders in a Duluth classroom, you might not want to donate your retirement savings to that project, no matter how well-raised a

Lutheran you are. An African mini-grid might, right now, return 5 or 6 percent on an investment, and as Roberts said to Banda, "You want to get them up in to the 10 to 12 percent range to attract more investment."

Numbers like that are truly frustrating. Let's take that $90 billion that Banda wants to build out Africa's mini-grids. As she points out, "That's, you know, a couple of billionaires knocked off. You know, one and two, probably not even one and two. Like, you know, position 19 and 20." (As it happens, the second half of the top 20 is a reasonable target: Rob and Alice Walton, along with brother Jim, are worth about $100 billion each, so . . . Africa plus an Asian country or three.) Climate campaigners from the Global South made the case at 2024's global climate talks that it would take a trillion dollars a year to really build out renewable power at speed around the planet. That's three Elon Musks, and in return you get a much cooler, much fairer planet. Or you could just put a 2 percent wealth tax on billionaires and that would get you $480 billion; add a teensy tax on financial transactions like buying stock and you'd get another $330 billion. As *The Guardian*'s climate editor Fiona Harvey pointed out, add in a modest tax on sales of technology, arms, and luxury fashion and you're pretty much where you need to be; or you could stop subsidies to the fossil fuel industry—that's about $270 billion a year in the West alone.

Justice certainly demands it. As Tom Athanasiou, the director of the activist think tank EcoEquity and perhaps the earth's most diligent environmental accountant, recently estimated, the US would have to cut its emissions 175 percent to make up for the climate damage it's already caused—since that's impossible, the only way to make up the shortfall is with money. As he told *The Nation* in November 2024, "True realism lies in the recognition that we actually have the money to save ourselves, and that the reallocation and redistribution of that money is now an existential necessity." The trillion dollars, he says, should come simply as grants—not as loans or investment flows, or anything else that keeps the poor world trapped in its current cycle of debt-fueled poverty. It would pay for the equipment—panels, windmills, batteries—that would free those people from the curse of buying the next boatload of coal or diesel. Mark Jacobson, using data from island countries in the Caribbean, estimated that the average consumer would see their energy cost fall to a fifth of what they pay currently if they could rely on their wind and sun, not on a tanker pulling up in the harbor. Imagine what that would mean for health care and education in Haiti or Jamaica. There's never been an easier way to cut the planet's ruinous inequality.

And yet none of that seems likely to happen right now—billionaire Trump won the White House with the help of Elon Musk, the world's richest man, and they seem far more intent on cutting taxes on the rich than on raising them. Trump blasted foreign aid on the stump (Kamala Harris, he said, "taxed money from the American taxpayer and sent it off to China and foreign regimes all over the world"). Last time he was in office he tried to make deep cuts in humanitarian aid, and this time, in an effort driven by Elon Musk, Trump chose USAID as the very first agency to shutter, trying to close it within two weeks of the inauguration. The only assistance that Trump's governing blueprint, Project 2025, envisions increasing for the Global South would be earmarked to help grow fossil fuels. So, while we wait and work for justice, there's other work to be done.

Much of it involves figuring out risk. One reason those Minnesota teachers would find it hard to invest their pension in Senegalese solar farms is that these are countries with a shaky grip on governance, given to rapid changes of policy at each election or coup, and prone to bouts of inflation or currency devaluation.

John Kerry, the former secretary of state who became Joe Biden's international climate negotiator, spent much of his time trying to find and deploy relatively small amounts of "concessionary capital"—grants from governments and charities—in order to unlock the much vaster quantities of private capital (all those pension funds). The idea is this: You deploy that government money through institutions like the IMF or the World Bank, using it to assume the risk of a currency collapse in Senegal or a nationalizing coup in Mali; and then those Minneapolis teachers could safely put their pension money to work. The last time we talked, Kerry put it like this: "If we just had tens of billions of dollars, we could leverage this transition very quickly and get people to make smarter choices." Tens of billions—pocket change to a single oligarch—could liberate trillions to save the planet: That's the definition of a bargain. A few weeks after talking to Kerry I was strolling the halls of the international climate talks in Egypt when I came across Sir Nicholas Stern—you'll recall him as the economist who provided the world with its best snapshot of the costs of climate change, nearly 20 years ago. Now a professor at the London School of Economics, here's how he put the dilemma: "With money at 6 or 7 percent, solar in Africa can outcompete anything and turn a profit. At 15 percent you can't make a profit, and so it won't happen."

I want to make two points here before I stop talking about money. The first is that we are so tantalizingly close. Always before we've seen environmental protection and economic development as at odds—now, all of a sudden, they're the same thing. Depending on how you calculate it, 10 solar panels per person is enough to give everyone American levels of energy. The Global South has 60 percent of the world's population but 70 percent of the world's renewable potential—remember, the solar power works better nearer the equator. Most developing countries are forever screwed in a fossil fuel world; India, for example, last year spent 5 percent of its GDP importing fossil fuels. But in a renewable world it can hold its own—more than hold its own. Most of the Global South doesn't have a giant legacy fossil fuel infrastructure, nor a powerful fossil fuel lobby dominating its politics; if it can get over the hurdle of capital, it's got a clear field to run. At the moment the sunniest countries have the least installed solar capacity. Germany, where no one ever took a beach vacation, has the most PV per capita, followed by Belgium and Austria; those are countries where good policy coincided with cash. But their lead won't last—the world is ready to be shaken up.

And the second point is that China looks increasingly likely to be the force that does that shaking—and if so, we will rewrite not just the physical basis of the planet but its geopolitics, as well.

I've been to China many times, but one of the most memorable trips was 15 years ago, to write a story for *National Geographic* on energy in the emerging superpower. The country was at the end of a decade of its most spectacular growth, most of it on the back of coal-fired generation; it had passed the United States three years earlier for the dubious honor of earth's biggest carbon emitter. But there were other signs too, and these already pointed to a different future.

I got on the train from Shanghai one morning to travel the short distance to the city of Wuxi. I climbed out into the most polluted air I'd ever breathed (and I've spent a lot of time in urban Asia). You could barely see across the street, but I managed the short walk to the headquarters of Suntech, which was already the biggest solar cell manufacturer in the world, though in 2010 that wasn't saying much. The building featured the largest building-integrated solar façade on earth—a huge sweeping wall of panels that would have been more impressive if the smog wasn't cutting their efficiency in half. But inside there was nothing but bustle—new employees were being added every day, and I sat in on an orientation session where they watched a dubbed version of Al Gore's documentary, *An Inconvenient Truth*. The wide-eyed young woman showing me around said she was proud to be there. "It's not only a job," she said. "I have . . . mission."

From Wuxi I traveled to Dezhou, a city near Beijing that was already billing itself as China's Solar Valley. The headquarters of Himin Solar dominated the landscape—the so-called Sun-Moon Mansion looked like a convention center surrounded by the rings of Saturn, great sweeping arrays of panels. Himin specialized in solar hot water heaters, which could be seen on the roofs of millions of Chinese homes and offices. Its CEO, Huang Ming, greeted me outside his office, in a private museum that held busts of his heroes—Michelangelo, Pericles, Sartre—and also one of the solar panels that Jimmy Carter had put on the roof of the White House and that Ronald Reagan had removed. His heaters were not especially high tech—essentially, black tubes that let the sun heat water. But he had drive—he'd been running revival-style marketing campaigns in cities across China. "We do road-showing, lecturing, PowerPointing," he said. And now Chinese tourists were coming to Dezhou to see the solar 4-D cinema, to ride the solar-powered giant Ferris wheel, to rent the solar-powered boats in the solar-powered marina.

Suntech and Himin may have been a bit too early—a few years later Suntech became the first Chinese company to default on American bonds when a glut of panels dropped prices, and as for Huang Ming, the market moved on; the one-time Sun King ended up in a series of disputes with Chinese authorities, looking back glumly on his glory days. "At that time, I was so hot-headed that I felt that I could do anything and earn so much money," he told the *Financial Times*. But in fact firms like these laid the groundwork for what came a few years later—the true explosion of Chinese renewable prowess. China, as of summer 2024, was building nearly two-thirds of the world's wind and solar projects, but it was also figuring out how to sell that tech around the world. Not in the US, where tariffs keep Chinese panels, not to mention Chinese EVs, safely at bay, but in most of the rest of the planet. We've already seen their solar panels popping up across Pakistan and Africa, but as NPR reported in November 2024, "Walk into an electric vehicle showroom in Colombia, the Dominican Republic, or Kenya these days, and the car on offer is likely made in China."

Yes, China continues to build coal-fired power plants—it has its own coal lobby, with powerful players—but many of those new plants operate half the time or less: It's cheaper to rely on the solar arrays that seem to spring up overnight. The air is far, far better; China has dropped pollution levels faster than any place on earth, and the ranks of the dirtiest cities in the world are now dominated by Bangladesh, India, and Pakistan. As a result, the average Chinese citizen can expect to live 2.2 years longer than they would have a decade ago, and one of the potential sources of trouble for the Communist Party has diminished. Meanwhile, strategic investments—China spent $329 billion on the clean technology supply chain between 2019 and 2023, according to Bloomberg, while the US and Europe spent $29 billion—have had strategic results. China, as Kingsmill Bond and his colleagues recently put, is the world's first electro-state.

The Inflation Reduction Act was America's effort to start catching up—in 2023, its first year of operation, 280 clean energy projects were announced, with $282 billion of investment; in 2022 China spent 26 times as much on the clean energy supply chain as Europe and the US combined, but in 2025 that advantage was expected to fall to just 2.4 times. As the Rocky Mountain Institute concluded, "That surge in capital expenditures is not enough to replace Chinese leadership, but it should be enough to provide [America] with much of its own clean technology demand." Or it would, if the US keeps it up—but Trump's 2024 election was like a jailbreak by an increasingly cornered oil industry. After candidate Trump, in a private meeting whose contents were soon public, asked the industry for a billion dollars in campaign funds, its grandees set to work—and as *The Washington Post* documented, no one worked harder than fracking baron Harold Hamm, who worked the phones for months, telling his fellow oilmen, "We've got to do this because it's the most important election in our lifetime." It worked—as one Trump aide said, "We've gotten max-out checks from people we've never gotten a dollar from before." The Inflation Reduction Act (IRA) won't be easy to scrap entirely (and the oil industry loves provisions that pay for boondoggles like carbon capture), but there's no doubt that much of America's momentum will be slowed in the crucial years ahead.

The question is how much that matters. As Trump removes America from the Paris accords, China will become the dominant player in international climate politics. *The Wall Street Journal* reported in November 2024 that "Beijing is using climate diplomacy to project its economic influence abroad in the developing world." At the last global climate talks it announced 2,000 "training fellowships" for scientists from around the developing world to help them better predict heatwaves, floods, and droughts, just as the Trump administration moved to close down similar decades-old programs at the National Oceanic and Atmospheric Administration. Since the developing world is where most of the demand for energy will come from in the years ahead, it's possible that having Trump's America on the sidelines may actually make progress easier in certain ways—countries as far-flung as Somalia, Cambodia, Sierra Leone, and Lebanon have seen solar panel imports from China in the last few years equivalent to half or more of their total electric grid. By 2035, the International Energy Agency reports, China's technology exports will be worth more than the oil exports from Saudi Arabia and the United Arab Emirates combined in 2024—China is, literally, the Saudi Arabia of sun.

In the best case, of course, the world would cooperate as it confronted the civilizational crisis that is global warming. (As those radicals at the *Financial Times* editorialized in early 2025, "China may be the leader in green tech, but the more important race is the one the planet is running, against the clock, to curb climate change.") It's possible, in fact, that the smartest course would simply be for everyone else to pay to run China's solar panel factories—instead of nationalizing an industry, you could globalize it. Those panel factories are currently operating at half capacity, but if you ran them around the clock you'd be producing as many panels as climate scientists say are required—about a terawatt a year. And then you could—well, you could just drop the panels on every wharf and pier and rail siding on the planet and walk away, letting people build out the new energy economy all by themselves.

Given political and economic reality, that's not going to happen—a more likely best case is that the US keeps up the momentum of the IRA and becomes a worthwhile competitor. But that may not happen either—given Trump's backwardness, it's entirely possible that the US could slide into a kind of global irrelevance, a slide greased by the incredible immaturity of its politics. Trump's first term as president, author Thomas Friedman wrote recently, was China's "Sputnik moment": He "lit a fire under Beijing to double down on its efforts to gain global supremacy in electric cars, robots, and rare materials." America invented the solar cell, and its university and government labs unraveled the mysteries of the greenhouse effect. The world will always remember Charles Keeling putting the first CO_2 monitor on the slope of Mauna Loa, a year before Hawaii became a state, and it will forever owe a debt of thanks to James Hansen of NASA, who became the Paul Revere of the climate crisis. But the US might decide to become an island of internal combustion, and then the essential nation might turn out to be China.

6

But Is There Enough Stuff?

The ugliest place I've ever been is the tar sands of Alberta, which is essentially Mordor. Some years ago, at the height of the battle over a project called the Keystone Pipeline, which had its origin in this hell, I joined Indigenous leaders from the area for what they called a "healing walk" across the region. For a very long day we hiked the roads north of Fort McMurray, trudging past vast "settling ponds" where the sludge from the mines combines with water in a toxic stew; any bird that landed on the surface would die, and so cannons fire day and night in an effort to scare them away. We covered the tiniest fraction of the war zone, which may be the single largest scar humans have ever inflicted on the planet's surface—you can spend bleak hours on Google Earth just scrolling across the endless miles of damage. My companion for part of the sad trek was Dr. Martin O'Connor, recently fired from his post as a doctor for causing "undue alarm" by pointing out the incidence of rare cancers amid this grim scene; he'd been showing me the "man camps," the dorms where miners stayed after they arrived for two-week stints in this barren outpost, and whose occupants have been regularly implicated in the abuse of local women. I'd been fighting the Keystone Pipeline mostly because of my fears for its effects on the climate—as the

NASA physicist James Hansen had explained, Canada's plan to quadruple output of the world's dirtiest oil would mean "game over for the climate." But now I had another set of reasons.

And that means that in turn I have no small sympathy for those who groan at the suggestion that a transition to using the power of the sun will require yet more mining. We need lithium, copper, rare earth minerals like neodymium—these are the elements that allow sunlight and wind to be translated into energy we can use. There is no way to make them appear magically; they will have to be gouged and scraped from the face of the earth. That gouging and scraping will produce tragic stories—indeed it already has. Kids have died in the "artisanal" cobalt mines of the Congo. As one miner told a reporter for National Public Radio, "Sometimes we are afraid because if you look at the ceiling [of the tunnel], you will see that it is already very fragile. The ceiling is already damaged. So, if we don't make repairs, at some point when you're down there, things are going to fall on you. And this can result in either a broken leg or a broken hand, or your skull will be fractured. Collapses are very frequent. We miners die a lot."

I think these fears—for the damage that mining does to places and to people—underly much of the antipathy to the green energy transition you find among progressives who are otherwise keen to tackle the climate crisis. And I think, as we shall see, that they need to be taken seriously. But it's often expressed in other ways, including with claims that we shouldn't even try to make such a transition because there's no way we'll find the necessary minerals. Andrew Nikiforuk, for instance, is a left-wing Canadian journalist that one publication described in a headline as a "nicer" version of Bill McKibben. "Neither mining nor technology are green or clean," he wrote; they will produce "more destroyed landscapes, debased watersheds, and displaced rural communities," and all for nought because "the global economy doesn't have the metals, rare earth minerals, energy, time or money to make this transition."

If this sounds somewhat familiar, it's because much of the rhetoric echoes the "peak oil" scare from 20 years ago, when a (noisy) school of thought emerged insisting that we were about to run out of fossil fuel, that we'd need to find 10 new Saudi Arabias by 2030, and so on. They convinced plenty of people. James Schlesinger, the retired director of the CIA and the Defense Department, declared "The battle is over, the oil peakists have won." But in fact, they were wrong—geologists discovered vast new oil fields, and they figured out how to frack existing wells to produce new flows; the world is awash in oil and gas. (That's why Trump, on day one, started permitting new LNG export terminals; the US, despite our huge consumption, has far more gas than we can use.) Humans are good at finding stuff, and the earth is a big place with a lot to find.

Which is why, for instance, the price of lithium, after soaring in the early part of this decade, has come sharply down in recent years—the quantity of lithium reserves and resources, which is to say known deposits the companies think they can mine, grew by a remarkable 52 percent between 2021 and 2024, with Argentina, the US, and Canada in the lead of new discoveries. A few months later the US Geological Survey reported that the US might have "all the lithium it needs in ancient brine which dates back to the Jurassic period and is buried deep below southern Arkansas" in what geologists call the Smackover formation. Meanwhile in California's Salton Sea, firms are figuring out how to distill lithium from the liquid brines of that infernal lake; the first extraction plant went online in 2024, and governor Gavin Newsom said that California might soon be the "Saudi Arabia of lithium." (Meanwhile in Saudi Arabia, chemists said in 2024 that they had learned to extract lithium from oil field runoff, and was constructing a commercial-scale pilot plant for what the Kingdom called white gold.) All of this reached peak craziness after Hurricane Helene that same autumn, when the internet was abuzz with the rumor that the federal government had directed the cyclone toward the hills of North Carolina in order to seize the town of Chimney Rock for a lithium mine. In other words, we're not going to run out of lithium, or graphite, or any of the other minerals that are useful for this transition; a 2023 paper in the journal *Joule* looked at 75 different scenarios for a green energy transition and ran the numbers for 15 different minerals. For most, demand through 2050 amounts to less than 15 percent of global reserves. The earth, the eight-person team concluded, "should suffice to meet anticipated demands." We may be a little short of tellurium, but I predict we'll find it.

And if we don't find it, I predict we'll figure out a way to use something else. Because that's always what's happened. A few years ago, for instance, a number of commentators said that the expansion of wind power might come to a screeching halt because of a shortage of the light balsa wood used for the cores of turbine blades—in fact, there was rapid deforestation underway in the parts of the Ecuadorian Amazon where the trees grew best. (One anti-wind group seized on the shortage to declare that the new power source did more damage than "a trillion Big Macs.") So turbine companies figured out a synthetic polymer foam for the blades: "Experts cite its higher temperature resistance and its comparative ease of recycling." It now accounts for more than half of new blades.

All of this is a way of saying that renewable energy relies less on resources than it does on brainpower; hence its footprint, while far from negligible, will be much, much smaller than extracting and burning fossil fuel—which, after all, is the comparison that counts. If you think about it for just a minute, you'll get the unnervingly simple point. Yes, you have to mine lithium to build a battery. But once you've mined it, that lithium sits patiently in the battery doing its job for a decade or two (after which, as we shall see, it can be recycled). If you mine coal, on the other hand, you immediately set it on fire—that's the point of coal. And then it's gone. And then you have to go mine some more.

Here's how Bloomberg did the math in 2023: "Annual demand for transition metals will grow fivefold by mid-century," they calculated. "Yet that doesn't mean we need to extract more stuff. In fact we need less. While EVs and clean energy infrastructure will mainly consume electricity and require lots of metal, the total amount of materials the world mines will fall."

How much? According to a large-scale report from the Energy Transitions Commission, "All the refined metals needed to reach net zero by 2050 will add up to less than the amount of coal mined in 2023 alone." Read that sentence again. Or look carefully at these numbers from *The Economist*: China's two biggest producers of polysilicon—the basic ingredient of a solar panel—each had a production capacity of 370,000 tons in 2023. Altogether, China has facilities capable of producing seven million tons of the stuff a year in its construction pipeline. Seven million tons of polysilicon, turned into solar panels and, when pointed at the sun, would add 3.5 terawatts of power to the world's supply each year, and they would go on collecting that sunshine for decades. Again, that's millions of tons of polysilicon. By contrast, the world mines eight *billion* tons of coal a year, and an equivalent weight of oil and gas. At the moment, something like 12 percent of the world's fossil fuel is used just to produce, refine, and transport fossil fuels, a job the sun and wind do for free. Over time the numbers become truly remarkable: If you ship someone a ton of coal, they can generate about two megawatt hours of electricity. If you ship them a ton of solar panels, over the subsequent quarter-century they'll generate about 1,000 megawatt hours—that's 500 times as much.

And those numbers will get steadily better, because we'll get steadily better at every part of the cycle—that's what a learning curve *is*. To give just one example: Rivian, the EV maker, was using a lot of copper wire in its vehicles. Then, in 2024, they announced a redesign—each car and truck would now need 1.6 *miles* less wiring. Companies announce new battery designs almost daily, and in every case they use less stuff, or—just as importantly—different stuff. Still worried about finding enough lithium? Good news: Novel batteries are emerging that rely on sodium instead. (They're already installed in some Chinese EVs.) And how much sodium is there? It's the sixth most common element on earth, with about 3.09 sextillion pounds. By some calculations if you extracted all the salt in the world's oceans you could form a crust 500 feet thick over all the earth's land. (We're actually more likely to use a mineable form called sodium ash as the foundation of the battery industry, and the US, for once, is out in front, with 92 percent of the world's supply—as *The Wall Street Journal* crowed, "The U.S. is the Saudi Arabia of the stuff.") Worried (as you should be) about that artisanal cobalt mining? New designs are substituting all sorts of things for cobalt—as Jenny Chase said in 2024, demand is lower than expected even as the supply has increased with the opening of new, better-regulated mines. You can tell what's happening by looking at prices. In 2024, even as the boom in solar and wind power accelerated around the world, the price of cobalt fell to a 10-year low.

If we're skilled at finding enough stuff to build out the clean energy transition, we're getting even better at the other end: figuring out how to recycle and reuse the stuff that's wearing out.

Truthfully, recycling solar panels and wind turbine blades is not a huge headache—not yet at least—because not many of them have yet worn out. That hasn't stopped clean energy opponents from making it an issue. A few minutes searching online will find you pictures of piles of discarded blades, and as *The New York Times* reported, these sites (there are three in the American West) "have a spooky nickname: wind turbine graveyards." (The spooky "fossil fuel graveyard," by contrast, is the entire atmosphere, which it's harder to take a picture of.) The same article, from the fall of 2024, reported that the solution is already in sight: a slight reformulation of the recipe for the foam in the blades, developed at the National Renewable Energy Laboratory, which utilizes "inedible sugar extracted from wood, plant remains, used cooking oil and agricultural waste" and can be easily recycled. In the meantime, old blades are being used for utility poles, or in bridge construction; they can also be "shredded into filler or added to cement production. A seven-ton blade that is ground and sent through a cement furnace can replace five tons of coal." (Even the nacelle, which is the technical name for the compartment that sits at the top of the turbine and houses the generator and the gearbox, can be recycled; the European wind giant Vattenfall is lowering decommissioned ones to the ground and refitting them as small but sturdy houses.) As for photovoltaic panels, as the Bloomberg analyst Jenny Chase wrote last year, "Sometimes the media gets extremely carried away with a particular 'dark side of solar' narrative." She cited a *Los Angeles Times* piece headlined "California Went Big on Rooftop Solar. Now That's a Problem for Landfills." Which, when you actually read the story, found that the state's landfills had been presented with 335 panels in 2021. Not 335 tons. 335 panels.

On the list of the world's environmental problems, then, dealing with cleantech waste is not *that* big. Especially compared with, say, fossil fuels. Coal and oil and gas can't be recycled because they're burned up, but they do leave behind carbon dioxide, which is melting the poles and burning the forests. And when you burn coal, you also leave behind some ash. Like, 110 million tons a year in the US alone, which is enough to fill about 1,000 coal ash pits. And talk about actual "spooky graveyards": One study found that 91 percent of these coal ash deposits were leaking toxic chemicals into nearby rivers and streams. Near Lake Michigan, for instance, researchers found a leaking coal ash pit that was producing arsenic levels two and a half times above the "maximum deemed safe by the federal government." A single coal plant in Texas, recently converted to a solar farm, was releasing 360 times as much cobalt as allowed by groundwater protection standards. The largest industrial spill in American history remains the 2008 blowout of a dam holding back a slurry of coal ash in Tennessee. A billion gallons poured out, swamping many homes; it took 900 workers five years to clean up the damage, and many of them got sick or died from what the Natural Resources Defense Council carefully described as "conditions that strongly suggested coal ash poisoning." These coal ash pits have been filling for a century or more, so they're essentially invisible to us, literally a part of the landscape (and usually in poor communities of color). But it sure makes turbine blade recycling seem like less of a problem. Indeed, by one reckoning a person who got all their electricity from wind over 20 years would produce about nine kilograms of blade waste; if they got their power from coal, they'd produce that much waste every six weeks.

Still, there's no reason to create even a small problem if you can avoid it, and with most of the detritus from the clean energy transition, that shouldn't be a problem. Think about batteries, for instance. There will soon be hundreds of millions of EVs on the road, and even if their batteries last far longer than originally anticipated, that still, eventually, will produce a problem. Or, actually, an opportunity: It turns out that even when an EV battery is old enough that the range of a car starts to drop noticeably, it's still got plenty of life, and so there's an emerging industry in taking them out of sedans and putting them next to solar farms. It happens first in the places that are poorest, and where recycling is a way of life. I quoted Tombo Banda earlier, describing the spread of solar mini-grids across Africa—many of them, she told David Roberts in a podcast interview, employ "second-life batteries." They were originally "used in dynamic settings, but now they can be used in stationary settings. We have some middlemen, who take these EV batteries, disaggregate them, look at the individual cells, make sure they are still performing well and still have sufficient life in them . . . and redeploy these to mini-grid sites," where they cut the cost of storage by about 40 percent. But the same thing's happening everywhere: There's an Irish company, for instance, called Range Therapy (the opposite of range anxiety) that is repurposing old EV batteries from "crashed and end-of-life vehicles" and selling them to homeowners. Instead of a Tesla Powerwall, Dubliners can buy a Range Wall or a Range Trailer, "a mobile clean energy source tailored for farms and businesses."

Eventually the batteries degrade enough that they can no longer be used as batteries and need to be recycled for their component parts, and luckily the stuff inside them—lithium, cobalt, and so on—is valuable enough that people have been hard at work figuring out how to make that recycling happen. The biggest American plant to do so opened in 2023 in Nevada; it's run by the former chief technology officer of Tesla, and it consists of a calciner, which is something like a kiln. In highly technical terms, you pour batteries in one end, and at the other end you collect their component parts—about 95 percent of the minerals come back, big sacks of lithium powder and piles of metal. (A company press release described it as, among other things, the first nickel "mine" to open in the US in a decade.) What does all this add up to? A 2024 report from the Rocky Mountain Institute predicted that by 2050 we'd have done *all the mining we'd ever need to do* for battery minerals; we'd just take them out of service and recycle them, over and over again. That seems like an unlikely claim—after all, we're only getting 95 percent of the minerals back—but remember that with each passing year we learn to build batteries with less lithium, less cobalt, less nickel; improving that efficiency by 6 to 10 percent a decade is enough to offset the recycling losses, and we're doing far better than that already. "Such a closed-loop supply system means we can continue to derive value from battery minerals for centuries. Over the next 20 years we will gather minerals not just to power the energy system of 2050 but also through to 2100 and beyond." Summarizing the Rocky Mountain Institute report, futurist Cory Doctorow put the 125 million tons of minerals we'll need between now and 2050 in context: "It is one-seventeenth of the amount of fossil fuels we dig up every year just for road transport. In other words, we're talking about spending the next thirty years extracting about 5.8 percent of the materials we currently pump and dig every year for our cars. Do that and we satisfy our battery needs more or less forever."

This combination of recycling and increasing efficiency makes for a kind of mind-blowing virtuous cycle. Consider the roof of my house. I was an early adopter of solar panels, putting them up beginning about a quarter-century ago when they were still expensive—but if someone spending his life fighting climate change wasn't going to spend the money, who was? I've added to those original panels three more times over the intervening decades, and each time they've been cheaper and more powerful. The original ones are still working fine (and indeed it was recently reported that Europe's oldest solar installation, atop a Swiss university lab, is still going strong after 35 years)—but when they need to be recycled, they will be like small mines themselves. In 2004, according to the Fraunhofer Institute, one watt of solar PV needed about 16 grams of polysilicon; this has dropped now to about two grams, or one-eighth the amount. As the Oxford data scientist Hannah Ritchie calculated recently, "The silver used in one solar panel built in 2010 would be enough for around five panels today." By 2035 or so, when my oldest panels go out of service, each one will almost certainly contain enough recyclable material for creating 10 new panels. I don't even know what to call this: super-recycling? It's more than circularity—it's like some kind of Möbius strip. It just gets better and better.

So yes, we can mine enough minerals for this transition. Which doesn't mean it's the *only* thing we should do. Andrew Nikiforuk, the nicer Canadian version of me who was insisting that we couldn't find enough minerals, also insisted that therefore we needed to make our lives smaller. Responsible leaders, he said, "would advocate for fewer cars altogether, shorter transportation networks and localized economies. They would make cities smaller and more walkable, and ban yachts, cruise ships, private jets, and SUV vehicles, whether with battery or combustion engines, because they represent a decadent waste of materials and energy." Which makes sense to me. Those are mostly things I've worked on. But they're hard—we've just fought and perhaps won a battle to stop the expansion of the main private jetport in Massachusetts, for instance, but it was in deep-blue Massachusetts, and all we stopped was the expansion—the original jetport is still there. So hopefully a clean energy transition will buy us some time to do these things.

And hopefully we won't have to take up that time with the sloppier ideas of other degrowth advocates: On X (Twitter), for instance, one regular in these debates, Aashis Joshi, argued that instead of each home having a washing machine "we could have communal washing facilities in each neighborhood" where people would gather in groups to wash their clothes by hand. "Washing clothes by hand is tiring work if you have a load," he admitted, "but it's still physical activity and exercise. We spend time in the gym and running outside to keep fit; would it be so bad to devote some of that time and energy to washing clothes by hand?" I don't know the right philosophical response to that question, but I was grateful to the energy blogger Jesse Peltan who tweeted back the observation that a washing machine uses 100 kilowatt hours of electricity a year, equivalent to the amount produced by something like one-tenth of one solar panel. I do my share of the laundry in our house, and that made me feel better.

A more difficult question than "Can we do this?" is "Can we do this fairly?" And there, I fear, the answer will always be more ambiguous. China opened its largest solar farm yet in June 2024—3.5 gigawatts, producing enough electricity to power all of Luxembourg or Cameroon or Laos (or Vermont or Alaska), one more sign that the country is peaking its carbon emissions years ahead of schedule. But it was built in what the Chinese call Xinjiang and the local Uyghur Muslims call East Turkestan—a colonized place. Local activists saw the solar farm as just one more way to extract value from their land, after oil and gas and cotton and indeed the silicon in many of the cells in the panels. "The solar plant is just the latest manifestation of those atrocities," one Uyghur activist told the website Atmos. "I believe anyone who praises China's pretentious commitment to green energy while failing to address the severe human rights abuses driving the industry, it amounts to complicity in the government's crimes."

I've never been to Xinjiang, but I have been to Tibet, where the Chinese are building the world's highest altitude solar farm, and where the ongoing repression of the Buddhist population is sad and sickening. And I've been to Inner Mongolia, on the great plains around the city of Ordos, where 5.9 million panels, enough to power two million households, came online in November 2024. This region is starkly beautiful, grassland steppe sweeping to the horizon—I highly recommend visiting Genghis Khan's grave, even if it's highly dubious that he's buried there. But this too is colonized land—in 2020, angry protests broke out over Beijing's decision to stop teaching in Mongolian because, as a government website explained, of the "inherent excellence of Chinese culture and advances to human civilization."

This, of course, is not a uniquely Chinese problem. American colonization is old enough that we no longer need to ban native languages—they were all but squashed long ago. And when Americans go looking for, say, lithium, they often land on contested soil. A big mine at Thacker Pass in Nevada, for instance, won the approval of one local tribe, but several others opposed it vociferously, on the grounds that it would wreck a sacred spot—indeed the site of a massacre of as many as 50 Native Americans in the mid-19th century. Indigenous groups appealed to the White House, but with little success. "We're kind of looking for some action. There hasn't been any, especially on the massacre site, which is a big issue to the tribes. We're also concerned about water and the environment and the cultural significance," one spokesperson told *The Guardian*. "When we talk about leaving something for our generations to come, we're trying to do that."

I think Indigenous groups have the right to say no to these kinds of developments, even where the laws are ambiguous—having been sacrificed in the last few industrial revolutions, it's right to let them exercise real sovereignty this time around. But I'm realist enough to know my opinion won't carry the day, not in the US, not in China, and not in lots of other places. There's real harm that will come to real places and real people as we build out this new energy future. And so for me the question becomes how that harm compares with the ongoing harm of our present system. Consider Inner Mongolia—the solar farm is not the first project that Beijing has foisted on the area. I spent a couple of days in the city of Ordos, the most absurd place I've ever been. It was the exemplar of China's real estate boom, a ghost city with hundreds of empty apartment towers, built on speculation and when I was there largely uninhabited. The massive central plaza is bigger than Tiananmen Square; and what I remember best about the plaza is the enormous library along one side. It was built to resemble three books leaning against each other on a shelf, and it has 2,300 seats for readers, 2,299 of which were empty after I sat down to write some notes. Carla Hajjar, an urban design expert in Shanghai, described the basic plan of the place: "The citizens have to feel really small compared to the city, they have to be afraid of the city." And not too far from the city one encounters some of China's biggest coalfields, with enormous machines clawing lignite from the ground like mad insatiable dinosaurs; I wanted to take pictures, but my local guide was too afraid of the police, and kept making me climb down in the footwell of our car.

What I'm trying to say is, colonization is an enormous sin. But I'm not sure it's made worse by solar farms or lithium mines, which at least offer the world some prospect of a break from the greatest colonization of all: the conquest of the air and the climate. Remember, nine million people a year die—one death in five—from breathing the particulates spewed out by fossil fuel combustion. Remember, every tenth of a degree Celsius that we raise the temperature moves another 100 million people out of a normal human climate, which has to be the greatest colonizing scheme of all time. We should work hard to temper the tragedies that come with every kind of extraction. But my guess is that stealing the sunlight of Inner Mongolia will, over time, do less damage than stealing the coal. My guess is that, for Indigenous communities that wanted such things, the advent of lithium mining offers them the chance to extract some serious concessions from the government, as Alaska's natives proved a generation ago when they took control of much of the state's fossil fuel territory. But these are only guesses, and this is not really how morality works. In a fair world, the people of Xinjiang and Inner Mongolia and Thacker Pass would get to make these decisions on their own, and decide for themselves what happens next. But we don't live in a fair world—we live in a world that's very rapidly tipping toward hell, a hell that will be hardest on precisely the people with the least power. Only moving fast can head off that hell, but moving fast means, inevitably, carelessness. It's a hard call; I'm so scared of the climate crisis that I may bend too far.

Which is one way of saying I'm lucky to live in Vermont, a place with almost nothing of any value beneath the soil—our only large-scale mining is granite, for tombstones and monuments. There are a few quarries in one small corner of the state, and I remember asking one of the executives of the mine how long they could keep digging. "At present rates?" he said. "About 5,000 years." We're not going to have to make any hard choices about digging stuff up where I live, or indeed most places. Which is not to say we're not going to have to make some hard choices.

7

Do We Have Enough Land?

The concentrated forms of energy—oil, gas, coal—are found only in a few places, and they're worth so much money that the riches to be made essentially determine politics. The machinery necessary to make the fossil fuel system work—the refineries and so on—can usually be relegated to places where opposition can be overpowered: That's why Cancer Alley lies amid poor communities of color in Louisiana. Those coal ash dumps I was describing a few pages ago? Seventy percent of them in the US are in low-income communities. The Dakota Access oil pipeline was originally routed near the North Dakota state capitol of Bismarck, until people there started to worry about their water supply—the obvious solution was to send it through an Indian reservation instead, which eventually produced the historic 2016 confrontation at Standing Rock.

By contrast, one of the things that makes sun and wind power so liberating—that they are available everywhere—also makes solar and wind farm projects relatively easy to stop. This is not an issue everywhere—Uyghurs and Tibetans don't really get to object. But in America, local opposition has emerged as one of the biggest obstacles to the rapid adoption of clean energy. A series of myths have been canonized on Facebook and YouTube, and they are used to rally opposition in place after place. By 2024, *USA Today* reported, "Clean energy plants are being banned faster than they're being built," with at least 15 percent of counties across the country "effectively halting" new construction through "outright bans, moratoriums, construction impediments and other conditions." Every year "we're seeing restrictions that are more severe," one expert explained. "You now have counties in Nebraska that have three-mile setbacks [from neighboring properties] for wind turbines, so if you have a square plot of land you would need 36 square miles to site a single wind turbine." In Wyoming, early in 2025, a legislator filed Senate Bill 92, the Make Carbon Dioxide Great Again act, that would outlaw any carbon reduction measure. The same week Oklahoma legislators rallied behind a bill to ban all renewable energy in the state, even though the Republican governor had just signed a deal with the Danes to collaborate on wind energy. (A potential political rival to the governor, the state's attorney general, denounced the deal because Denmark was "quasi-socialist.") It's not just the US—Rishi Sunak and Liz Truss, the last two conservative prime ministers in the UK, warned of solar panels "filling" England's farmland, though solar panels currently occupy 0.1 percent of that green and pleasant isle; the Tory health and energy minister Matt Hancock fought one proposal on the extremely Trumpian grounds that a "local golf course was at risk."

Since solar panels and wind turbines are not by themselves dangerous—they don't give off clouds of smoke or blacken streams—opponents have had to concoct a series of fantastical arguments against them. Donald Trump's infamous claim that wind power causes cancer, for instance, is a distillation of an internet theory that the noise from the spinning blades is causing everything from autism to depression, sometimes with links to reports about Portuguese horses; every medical study refutes any connection (one Canadian study found that "a small minority of those exposed express annoyance," but added "annoyance is not a disease").

Often, it's the fossil fuel industry that's helping spread this misinformation. In Ohio, for instance, "questions emerged about the funding" behind an anti-solar group after it hosted "a town hall meeting at the local theater with complimentary food and drinks" for 500 attendees. The questions were answered when it turned out that "a well-connected natural gas executive is among the group's largest donors." (The executive said he was unaware that the pictures he was spreading of damaged solar panels came from the island of St. Croix after the passage of a Category 5 hurricane, which a reporter, Kathiann Kowalski, noted was "a highly unlikely event in central Ohio.") In Michigan, a similar group sprang up to oppose renewable energy projects; though it described itself as "a pure grassroots coalition," its website showed that several of its key position papers had been written by employees of a GOP-connected consulting firm that also represents pipeline companies. Along the New Jersey coast, as *The Wall Street Journal* reported, "anti-wind activists suddenly fell in love with the right whale," using the endangered cetacean as a prop in their campaign against offshore wind farms. The Save Right Whales Coalition "grew rapidly," routing donations to the fight through a nonprofit called Civilization Works with close ties to the Koch brothers, America's biggest fossil fuel barons. (The leading cause of whale deaths, it's worth noting, is being hit by ships, and 40 percent of all maritime cargo on this planet is coal, oil, and gas; one container ship full of EVs will avoid the need for 84 tankers full of oil.)

Sometimes this kind of lobbying is truly cynical, drawing on our deepest divisions. In Maryland, for instance, something called the Energy Poverty Awareness Center was pushing the legislature to gut climate legislation on the grounds that it would harm communities of color. As *The Washington Post* dryly noted in January 2025, the former NFL star leading the effort "didn't mention that his group has ties to the natural gas industry." The vice president of the Energy Poverty Awareness Center was also an official at the American Fuel & Petrochemical Manufacturers, a DC lobby group. As actual advocates for the community told the *Post*, "These front groups are a way to obscure the harms the industry causes these communities through increasing pollution and exacerbating climate change, which hits disadvantaged people the hardest."

The idea, however, that this kind of opposition is confined to right-wing friends of the fossil fuel industry is simply and sadly wrong. Liberals don't like change, either, which is why my deep blue state of Vermont has had a de facto moratorium on wind power for more than a decade. In 2021 its Public Service Commission, well-stocked with progressive Democrats, rejected a 10-acre solar farm purely on aesthetic grounds. The developer had proven it wouldn't affect the rare Indiana bats or wood turtles because the only roost tree would remain standing and because there would be "wood turtle entry and exit holes" in the fence. They'd hired an expert to ensure that no archaeological remains were on the site, and another to certify that it wouldn't endanger the public health. But the commission noted that the trees planted to screen the view would take "several years" to grow in and that was enough: The project would be "offensive or shocking to the average person," the commissioners decided, and so they said no. In Santa Fe, New Mexico, opposition from local residents is delaying plans for a much larger solar farm, one that would power 37,000 homes in a town with about 50,000 dwellings. Santa Fe voted 3–1 for Kamala Harris in the 2024 election, but never mind—the opponents have built out a full-scale campaign centered on the idea that the batteries in solar farms present a fire hazard, even though the developers pledge to install the most advanced fire suppression system, and the nearest house is a mile and a half away. (The real fire risk in Santa Fe? Not surprisingly, it's climate change: In 2022, amid record temperatures, nearly a million acres of the state burned, thousands of residents had to evacuate, and the smoke plume sent many to the hospital because they were unable to breathe.) None of that prevented the citizens group from piously paraphrasing the British statesman Edmund Burke—"All it takes for bad things to happen is for good people to do nothing," their website declared.

That kind of smugness permeates these fights—sometimes I think the state motto of my beloved Vermont ("Freedom and Unity") should be amended to "Change Anything You Want Once I'm Dead." Many of us, especially perhaps those in rural areas, are used to the way the landscape looks right now, and even the obvious danger of climate change and the obvious appeal of power from the sun can't override our knee-jerk reaction: "I don't want to look at that." A solar farm or a wind turbine reads in our mind as "industrial," and since we have a deeply pastoral sense of the landscape, even if we're making our living from staring at screens, that strikes us as obnoxious. (I recently and joyfully welcomed my first grandson, and so I've been reacquainted with the fact that we raise our infants as if they will all be small farmers—"What sound does a sheep make?") But all this obscures several powerful truths about our landscape.

One truth is, we actually don't need very much land to provide the energy we need. At the moment, according to Stanford's Mark Jacobson, fossil fuel infrastructure takes up about 1.3 percent of America's land area—this includes active and abandoned oil and gas wells and coal mines (since, unlike the sun, these play out, you need new ones every year) and deforested strips for pipelines, power plants, and tank farms. By his calculation, converting entirely to clean energy would use less of the landscape. It depends on how you count it, of course—when Jacobson looks at the acreage of a wind farm, for instance, he includes just the pads for mounting the turbines and the paved roads between them, since everything else can still be farmed. His numbers, across 145 countries: "The total new land area for footprint required . . . is about 0.17 percent" of their territory. By contrast, at the moment the US devotes about 41 percent of its land—both pasture and cropland—to feeding cows. We devote two million acres to golf courses and three million to airports.

We can and should produce a good deal of our energy from rooftop solar panels and solar canopies over parking lots—but there aren't enough of them to produce everything we need, and it's considerably cheaper to use cleared land. Like, for example, some of the fields where we currently grow corn, the most widespread crop in America. And here—as someone who lives in a corn-growing county—is where I want to make an argument that may seem at first blush unlikely: *Converting some of these fields to solar panels makes enormous ecological sense.*

That's because one way to look at a field of corn (or any other crop) is that it's already an array of solar panels. A plant is a way to convert sunshine into energy through photosynthesis, which is an enormous miracle—the chlorophyll in the leaves absorbs energy from the red and blue parts of the spectrum, which energizes electrons, moving them to a higher energy state. A miracle—but not a very efficient one. Somewhere between 1 and 3 percent of the sunlight falling on a leaf actually becomes energy. The photovoltaic panel works considerably better: As we've seen, the average panel is about 20 percent efficient, and we're on a course that might someday soon get us to 40 percent efficient. Which means that, say, if you want to use corn to power a car, it takes a lot of it. About 40 percent of America's corn crop is turned to ethanol—in Iowa, on the richest topsoil in the world, that number is over 60 percent. If you spent a day driving past Midwestern corn fields, mostly you'd be seeing gasoline plants. But, again, inefficient ones: A few years ago, 200 scientists at 31 colleges and universities across Iowa signed a statement noting that a "one-acre solar farm produces as much energy as 100 acres of ethanol." Or, to do the math in reverse, an acre of corn will produce enough ethanol every year to drive a Ford F-150 pickup about 25,000 miles. But cover that same acre in solar panels and you will produce enough juice to drive the electric version of the same truck—the F-150 Lightning—about 750,000 miles. Or to do the math one more way, you could supply *all the energy* the US currently uses by covering 30 million acres with solar panels. How much land do we currently devote to growing corn ethanol? About 30 million acres.

And a field of corn is not just an inefficient solar array, it's an inefficient solar array on to which you have to pour huge quantities of nitrogen and phosphorus every year to make it work, nutrients that wash downstream where they cause endless trouble. Art Cullen, editor of the *Storm Lake Times Pilot* in Iowa, last year described for *The Washington Post* a journey in one of the electric Ford pickups from his home down to New Orleans. They began the drive along the banks of the Raccoon River, which flows into the Des Moines, which flows into the Mississippi. "When we began, freshly planted Iowa river bottoms, enriched by fertilizer, lay inundated by heavy rains that broke the back of a four-year drought here in the Corn Belt. About 30 percent of the nitrogen applied for raising corn is lost to water, and much of it right now is draining off in the spring rise." It all eventually reaches the Gulf of Mexico, where it forms a giant "dead zone," now roughly the size of New Jersey, a place so deprived of oxygen that fish can't survive. (Perhaps we really *should* call it the Gulf of America.) That Gulf coast, of course, is also home to many of the fracking wells that produce the gas that gets used to produce the anhydrous ammonia used as fertilizer, which then washes down the Mississippi again. Oh, and all the time it's getting hotter, raising the sea level around New Orleans.

Let me bring this back to Vermont. In 2024, a developer proposed a 300-acre solar array in a town in my small county; if built, it would be the biggest in the state, covering about 2 percent of the town's 14,000 acres, land currently used to grow corn. The owner, part of a farming family, said the lease was the best way to guarantee the ongoing medical care of his aged mother, but the reaction from his neighbors was immediate and blunt—within a week, 100 of the town's roughly 700 residents had signed a petition insisting the project would "degrade the rural character of the area," "contribute to blight," and "be antithetic to the landscapes that make Vermont special." As one resident said, he had moved to the town "not to look out the back door and see a field of solar panels," but "for the rural environment." It was true, some said, that the plan would generate substantial tax revenue for a town that had just seen its annual budget jump by 28 percent, but that was insufficient to quell the outrage. In the November 2024 elections Panton voted 2–1 for Harris over Trump, and 3–1 against the solar farm—which may not settle the issue, since the state public service commission will eventually rule.

When they do, they might want to take into account a couple of other facts. For one, the 300 acres are currently used to grow corn and other animal feed for Vermont's dairy industry, which produces more milk than anyone needs; as a result, prices are low, which puts economic pressure on dairy farmers, and 90 percent of the state's dairy farms have gone out of business since World War II. In order to grow the feed for cattle, the current farmer spreads 250 pounds of the herbicide glyphosate across the land and adds 300 pounds of dry fertilizer—nitrogen, potassium, and phosphate—*per acre*. Some of that almost certainly leaks into nearby Lake Champlain—as another Panton farmer, sued for his pollution, explained to a state court: "How are you supposed to stop 300 acres of water from going back where it naturally went in the first place?" The farm runoff pouring into the lake is a major cause of repeated algae blooms that close beaches and kill pets; the latest estimate, a few years old, of the cost of cleaning up the lake was $2.3 billion. Meanwhile, climate-driven floods are, year after year, causing immense damage across this small county.

All of which is to say, maybe we should think of clean electrons as another crop to be harvested. A crop valuable enough to return considerable tax revenue to the local community, a crop that contributes more to the common good than animal feed, a crop that does far less damage than the ones we're used to. Did I mention that Vermont pays $2 billion a year for fossil fuels? Money that heads to Texas and Saudi Arabia when we could be producing the energy we need right here, and running it through heat pumps and EVs. You don't have to cover all, or most, or even much of the landscape with solar panels. But we need to cover some, and a slight shift in our way of seeing would help. Yes, there's steel and glass in the field, but it's no more industrial than the chemical soup that is a cornfield; it's just a different way to catch the sun.

Happily, there are places where it's possible to see such shifts in attitude underway; I had a fairly good view of the future from a booth in the Boone County Family Diner (two eggs, bacon, toast, and potatoes for $6) in Poplar Grove, Illinois. Out the window and across the two-lane highway I could see a cornfield growing ethanol, and in one corner of that field an array of solar panels.

The array covered 36 acres, providing enough power for about 800 homes, and it paid the town about $11,000 a year in property taxes, compared with the $400 it produced when it was still in corn. Illinois—with deep blue Chicago driving its politics—has permissive solar-siting laws, but even so it takes finesse and care to persuade local communities to go along. I was sitting with Jon Carson, the founder of Trajectory Energy Partners, and Hal Sprague, his head of community relations. Their company had developed a number of these medium-scale projects across Illinois, and I was curious how they'd gotten it done. The answer, it turned out, involved going to a lot of hearings and listening to a lot of objections. Some, said Carson, are "straight up NIMBYs. And some of the worst are progressives—they always begin by saying 'I'm a big supporter of renewables, but not here.'"

"'I moved out from the city to look at corn,'" says Sprague, repeating another objection, and who had clearly heard this line too often.

To deal with these objections, the Trajectory team tries to talk with every neighbor. "On this project we had one guy we could never find at home," says Carson. "And when we had the zoning hearing there was one guy who showed up that we didn't know, and I figured it must be him. I was sweating what he'd say. And he said, 'Ever since I moved in, I've wondered what they were going to do with that land. I'm so glad it's solar, not a Lowe's distribution center with lights all night and trucks backing up all day.'"

But they also work hard to find allies. Some are farmers. "One of these solar arrays might be 40 acres out of 4,000, which is about what you need to farm out here. So it's a nice hedge. You don't need any fertilizer, and you don't have to worry about the price of corn," Sprague said.

"It's not about the message, it's about the messenger," said Carson. "If the person who owns the land is from the community and they've decided to do it, that trumps a lot. And a landowner who's been part of a different project that's already built—that's huge."

And there are plenty of other supporters, too. The nearest city to Poplar Grove is Rockford, a Midwestern industrial center that's clearly seen better days. "We were the toolbox of the country," one local told me as we sat at Katie's Cup, a coffee bar. "Machine tools, smelting, metallurgy—during the Cuban missile crisis people said we were one of the top three targets for the Russians. Washington, New York, and Rockford." But the deindustrialization of the 1980s ended all that; before long the unemployment rate was 25 percent. Now the city is slowly recovering, and one point of pride was the solar farms popping up on vacant land around town.

I wandered over to Ernie's Midtown Pub for lunch (Wings Wednesday!) with another group of Rockford boosters. One of them, a Baptist preacher named Jeremiah Griffin, offered grace, and then they went around the table explaining what solar meant to them.

Brad Long, the head of Carpenters Local 792 (and lead guitar for Long Shot, which plays outlaw country and acoustic blues), said all the community solar farms are built with union labor—indeed, wages are about 60 percent of the cost of building one. "And it's been a great thing for new apprentices—the young people are actually interested." Keri Asevedo, head of the local Habitat for Humanity chapter, said they'd been building all-electric homes, complete with EV chargers, heat pumps, and induction cooktops. "People say, '*We* get *that*?' They're humbled, honored to be helping the planet." Mike Gallagher, a community organizer, ran the numbers for his local parish when the boiler broke, arguing that solar panels on the roof would save on the $85,000 cost. Tretara Flowers started a nonprofit called Reconnect 815 when she got out of prison in 2005. "I realized there's no one to help reentry individuals with interview clothes—with anything, really," she said. Now many of the ex-cons she works with are training as solar installers. "When a man builds something—well, they're eager to tell their friends what they've done. You feel a sense of ownership when your work impacts the entire world."

And then there was Griffin, the Baptist pastor. He's got nine kids, many of them foster children, and he lives in one of the grittier parts of town. I had the distinct impression there are things I would disagree with Griffin about, like who should be president, or the role of guns in our society. But I liked him enormously, and could tell he was a Christian in the actual sense of the word, deeply devoted to his neighbors. And man, did he like solar power. In fact, after lunch we adjourned to the former Rockford landfill, which Trajectory developed into a 10-acre solar farm. "This was a quarry where people would just come and push whatever they wanted in," he said, as we took in the sweeping view from this high point in the city. "And it was in the middle of a gang shooting at the time. I'd come up here and pray for the city. And now to be here is the exact opposite."

So far I've been describing a choice between producing crops and producing energy. But in lots of places there's an emerging middle ground. The word *agrovoltaics* was coined in 1982 to describe solar farms that are also farm farms. From a small start—say, grazing a few sheep to keep the weeds from shading the panels—it's become a serious part of the clean energy future. A recent survey found 72 percent of German farmers considering the deployment of agrovoltaic arrays on their land, and the French energy giant TotalEnergies recently set up a Center for Expertise in the art of growing food next to PV panels—and no wonder since a 2024 study found that the presence of solar panels can increase yields for Chardonnay grapes by as much as 60 percent. There are a lot of studies like that, and they serve as a reminder that for many crops some shade and some extra humidity are a blessing, especially as temperatures rise. Near Phoenix, temperatures topped 110 degrees for 31 straight days in the summer of 2024, smashing old records. "The solar arrays help reduce our water use," a USDA agronomist told a local reporter. "Plants don't really need as much sun as they get in the West." Even skipping irrigation every other day, soil moisture under the panels was 15 percent higher than nearby unshaded plots, and black-eyed peas were growing faster because they were less stressed.

Researchers have found much the same in Africa—test plots in Tanzania and Kenya, for instance, revealed that corn, Swiss chard, and beans were thriving under panels. "We created a microclimate that helped certain crops produce more, but they were also better able to survive heat waves and the shade helped conserve water, which is crucial in a region threatened by climate change," the study's lead author explained. Not only that, the solar panels produced more than half the power used on the farm. As an Englishman putting a 100-acre array on his Cambridgeshire farm explained to reporters, "It's not 'produce ten units of energy' or 'produce ten units of food.' It could be six units of both. And then, all of a sudden, your two halves are greater than the whole." Indeed: Australian trials found that wool from merino sheep improved in both quality and quantity on farms with solar panels—as the *Independent* newspaper explained, the panels provided "shelter for the sheep and the grass." A few moments on Google will provide you with endless pictures of livestock relaxing in the shade of solar arrays, though the emerging wisdom is keep them away from goats—they tend to eat cables, and they're liable to try and jump on the panels. Cattle, though, can work—a large-scale French trial outfitted bovines with sensors "to measure activities such as ingestion, rumination, rest, and standing," and "the presence of the panels does not seem to modify their activity." Cows just carry on.

It's even possible to imagine this kind of scheme working on the vast grain farms of the Midwest. Ohio State researchers are currently trialing panels amidst alfalfa, hay, and soybeans. The giant combines could pose a problem, but as one researcher said, "I could get my 20-foot seed drill in there, I just have to be careful."

Even when you're not growing food or grazing animals, though, there's increasing evidence that solar arrays can help the land where they're installed. In Phoenix, surrounded by desert, windstorms increasingly blow dust hundreds of feet into the air—that's because the crust of the desert's soil is so easily broken, even by a footstep. But this "biocrust" seems to be naturally regrowing under solar panels, which researchers at Arizona State described as "beach umbrellas." Indeed, they're now growing these crusts under solar arrays to replant in damaged desert areas; "crustivoltaics," they're calling it. In China, where great dust storms have swept off the Gobi into the cities of the north for decades, they're finding the same thing: The ground under solar arrays had higher plant and microbial diversity than the surrounding desert. Indeed the arrays themselves were acting as windbreaks, keeping the dust out of the air.

It's useful to remember that, as bad as the climate crisis is, it's not the only ecological dilemma on the planet—when you talk to scientists, the loss of biodiversity is the other disaster that drives them to despair. Photovoltaic energy can help here, and not just by lowering temperatures. An April 2025 UK study found triple the number of birds on "mixed-habitat" solar fields as on adjoining cropland. And as with birds, so with bees.

I'm going to end this chapter with a story about my favorite field trip of 2024. It didn't take me far—maybe a hundred miles north and west across Vermont—but it left me with a very different sense of how the world might work.

It was an August morning, and Tawnya and Mike Kiernan drove me north on Route 100 toward the town of Stowe, talking nonstop. We pulled off the two-lane highway and onto a short farm road, and then got out at an access gate along a wire fence that enclosed an 11-acre field of solar panels. The reason we were there is that three years ago, Encore Renewable Energy, a Burlington-based developer of solar arrays, contracted with a nonprofit that the Kiernans started, called Bee the Change, to seed pollinator-attracting plants that are native to the area in the rows between the panels. The organization's small crew tends more than 20 fields like this across the state, weeding and, at least once a year, mowing what they have planted so that it doesn't grow so high it shades the panels.

The approach seems to be working. When the Kiernans are hired by a solar developer, it's usually to plant on what was until recently a farm field; because the fields are typically monoculture and have been treated with pesticides for years, "the pollinator density is really low." Mike uses a pollinator-counting method that involves walking on the margin of a field and counting wasps and flies and moths for seven and a half minutes. Then a random-number generator tells him which row of solar panels to walk along, and as he walks he counts the pollinators he sees in seven and a half minutes, then adds the two numbers together. "On those abandoned farm fields, we might get a count of forty or fifty in fifteen minutes," Mike said. "But now, once we've done our thing, you can see ten at a glance. You can see 300 in 15 minutes. You see a lot of them even this late in summer, during what we call a 'dearth period.' Wait till next month, when the asters come in!" We saw, indeed, an extraordinary number of small insects flying in and out of whatever blossoms were available: goldenrod, mountain mint, evening primrose, black-eyed Susan. "That echinacea is pretty tired," Mike noted, then pointed out a lacy thing resting on a fleabane blossom. "Here's an ichneumon wasp," he said. "It lays its egg inside the larvae of other insects—there are 40,000 identified species in this family of wasps. Here's a fly in the family globetail—enormously important as a pollinator. That's from the housefly family, a muscid. On a cold day like this, it's mostly flies." (After bees, flies are considered to be the most important pollinators.) While we spoke, Tawnya never stopped pulling weeds that the Kiernans don't want in these fields, such as the aggressive yellow foxtail, an invasive plant that reestablishes itself when equipment is brought to mow or service the fields, carrying the seed with it. "Or look at this ragweed—it could have 30,000 babies," she said, ripping it from the ground. "You just have to keep after it," Mike said. "There are maybe thirty species that are problematic."

The Kiernans founded Bee the Change about a decade ago, when their kids left for college. (Mike is also an ER doctor in Middlebury, and Tawnya is a pediatrician, but their empty nest, they insist, left them with "no idea what to do with ourselves.") They were interested in pollination, in part, because of time that Mike spent as a volunteer doctor in Haiti, a country where widespread deforestation and ongoing damage from strong hurricanes has dramatically depleted the number of pollinators. He observed that roughly a third of the patients he saw were nutrient deficient. "People are missing the B vitamins from fruit and vegetables. If you want to see a world without pollinators, that's it." As the nonprofit's name implies, their first tools were honeybees; they installed hives in solar fields. But the more they learned about biodiversity, the more they wondered whether this strategy was actually the best for the environment. Honeybees are domesticated and are so persistent and numerous—more than 30,000 can live in one hive—that, in Mike's words, they "can put too much harvesting pressure" on the plants. There may not be enough nectar left behind for all the wild pollinators, a complication that spells peril not just for them but for the plants they're particularly adapted to. "There are more than 350 *native* bee species in Vermont," Tawnya said. So they stopped placing hives and started installing native plants that attract wild insects.

By now, our tour had traipsed through a former auto junkyard turned solar farm, where we looked at boneset and lots of milkweed. Tawnya harvests milkweed floss to produce what she calls an excellent down alternative, and uses it in several of the products available on the Bee the Change website. "I make neck warmers with it, and they're almost too hot, except on really cold days," she said. Then we drove south back down Route 100, eventually arriving in the picturesque town of Warren, where the town's school playing fields are backed by several long rows of solar panels.

The Kiernans have been tending this particular array for a couple of years. The sun had come out now, and the early afternoon temperature warmed to near 70 degrees, sound conditions for luring insects into the air. "That's *Halictus ligatus!*" Mike said, as we walked between the panels, excited even though the ligated furrow bee is a relatively common wild bee. Then, sounding truly elated, he announced, "That's *Triepeolus pectoralis*, a native, but one we rarely see! That's only the tenth time it's been seen in New England." Next he was on his knees beside a particularly vibrant swamp milkweed. "That bluish-black wasp is a mud dauber," he said. "He can catch a spider and keep it up on your rafters with a little bit of mud." Pulling out his phone to take a picture, he added, "This bee with the red on his abdomen I've never seen before. Sometimes you need to get the male genitalia under a microscope to make an identification. You wouldn't think so, but it's really, really species-specific."

We watched an orange-collared moth, and then a type of fly that grabs onto bumble bees in mid-flight and lays eggs in their abdomens. "And here's the great golden digger wasp," Mike said. "It's extraordinary to look at and virtually harmless. And that's a bee fly. Her lifestyle is unbelievable. She lays eggs in the holes of ground-nesting bees by dropping them in like bombs. *Bombylius*."

All the while, the solar panels stood quietly by, converting the late afternoon sun into clean energy, energy that might help spare the future for the kids playing soccer nearby.

SECTION THREE

LET'S DO THIS!

8

Time to Push

I went to work for the weekly newspaper in my Boston suburb when I was 14, writing sports but also increasingly covering what passed for news in the well-ordered precincts of Lexington, Massachusetts. This was the 1970s, and the first truly significant world story I got to play a part in involved the twin oil shocks of that decade, when OPEC cut the flow of crude to the US and prices skyrocketed. Of course, my "part" in the story mostly involved calling up locals: Al at Gulf and Dom at Amoco and asking what hours they'd be pumping. ("Hours? I have no hours. I pump gas when I feel like it," Dom snapped.) Or I'd ride my bike over to interview people as they waited in line; by the time the second of these crises happened in 1978 I had my driver's license, so I could take my family's maroon Plymouth Fury and wait in line myself.

I tell this story because I want to make a point about the contingency of history. We came very close back when I was young—far closer than I really understood in those years—to making the crucial changes we still need to make. Now, 50 years later, I'm old, and we're in much the same place. And I very much fear that, without a burst of inspired activism, we'll blink again and turn away. We won't abandon energy from the sun—there's too much momentum—but we'll slow our pace, and as a consequence the world will fatally overheat. We need to do this *now*; we can't afford another miss.

Before the energy crisis—even before the solar cell was invented—President Truman had commissioned the owner of CBS, William S. Paley, to chair a panel to look at postwar resource needs of the nation. That group's 1952 report included a data-filled chapter on all the possibilities for energy from the sun. They looked at the numbers from the wind turbine on Grandpa's Knob in Vermont, they investigated early models of electric heat pumps, they examined early versions of what we'd now call passive solar houses, and they concluded that "if we are to avoid the risks of seriously increased real unit costs of energy in the United States, then new low-cost sources should be made ready to pick up some of the load by 1975. . . . We must look to solar energy."

That report produced no action, however—we were, after the war, in a period of great centralization, building big things (the new energy that excited people was mostly nuclear, at least until the costs began to mount). Solar power perked along—the new solar cells worked great for satellites, where cost was not an issue, and Western Electric patented a dollar bill changer that ran on solar power. Offshore oil rigs, lighthouses, railroad crossing signals, scientific stations in polar regions, and then calculators, and wristwatches. It wasn't until the first oil crisis in 1973 that people started taking it more seriously—that year, for instance, researchers at the University of Delaware built Solar One, the first house heated and powered entirely by solar energy. The 1,300 square-foot house (which still stands, on the corner of South Chapel Street in Newark, Delaware) drew 100,000 visitors in its first year.

It wasn't just the energy crisis; there was also this new thing, "environmentalism," that had been born with Rachel Carson's *Silent Spring* in the early 1960s and which exploded at the first Earth Day in 1970. People were angered by leaking oil rigs and burning rivers, and intrigued by the first images of our earth making their way back from the Apollo missions. In 1972 a mammoth bestseller called *Limits to Growth* insisted that we couldn't simply keep expanding our economy without wrecking the earth. Solar power seemed a way out of this conundrum—a way to generate energy on a scale we needed without pollution, a way to free ourselves from dependence on Saudi Arabia. There were hippies involved—but hippie was not such a weird thing at that moment in America (Jimmy Carter had Willie Nelson over for a visit to the White House, which ended up with the singer and an unnamed accomplice smoking weed on the roof; Carter also held a reception for E. F. Schumacher, whose argument for "Buddhist economics," *Small Is Beautiful*, was atop the bestseller lists). And there were also all kinds of serious national security people engaged. Back in Lexington, I got to interview Carter's undersecretary for energy, an MIT professor named John Deutch. "If you're a person in a situation where there is too little oil, you would look for easy solutions and scapegoats," he told me, as he sat barefoot and in jeans on his back porch. (Again, this was the 1970s.) "But there is no quick fix."

Instead, Carter decided, there was solar. As the decade had spun along, all kinds of people had gone to work on the project, advocates and engineers and advocate-engineers. After months of looking, I finally found (buried in the stacks at the Boston Public Library) a copy of a book called *Sun: A Handbook for the Solar Decade*, that came out in 1978 with a foreword by legendary environmentalist David Brower ("people are sorely needed now to work the solar side of the street. . . . The opportunity is real, immediate, and not likely to be offered again"), and essays by

Earth Day founder Denis Hayes, Amory Lovins (who went on to found the Rocky Mountain Institute, whose analyses you've read often in this book), Ivan Illich, and Lewis Mumford. It's a staggering volume, filled with technical precision, political shrewdness, hopeful naivete. And all these efforts worked—they convinced the White House that solar energy was the way forward. Carter, in 1978, announced the first Sun Day, traveling to the federal government's mountaintop research facility in Golden, Colorado, to give a remarkable speech that summed up the decade's various strands. "Nobody can embargo sunlight," he said. "No cartel controls the sun. Its energy will not run out. It will not pollute the air; it will not poison our waters." Carter—with characteristic bad luck—was giving this speech outside in a driving rainstorm, not the backdrop his handlers had hoped for. But he was resolute. "The question is no longer whether solar energy works," he said. "We know it works. The only question is how to cut costs."

And the way to cut costs was to spend money. Carter's proposed 1980 budget called for a billion-dollar Solar Bank to fund research and offer loans to homeowners who wanted to put solar panels on their roofs. The point of his plan, he said publicly, was that by the year 2000 a fifth of the country's electricity should come from solar power. That was clearly the timeline he had in mind in 1979 when he climbed atop the White House, up where Willie Nelson had smoked his joint, to inaugurate its first solar panels. "A generation from now," he said, these will either be "a curiosity, a museum piece, an example of a road not taken, or it can be a small part of one of the greatest adventures ever undertaken by the American people." That same year John Hall released his song "Power": "Give me the warm power of the sun," he sang, with Carly Simon chiming in. "There's so much to gain and so much to lose."

We made that choice in 1980, of course, electing Ronald Reagan in a contest not unlike that of 2024: A decent leader trying to look to the future was beaten by a say-what-you-want-to-hear demagogue, mostly because people were worried about inflation. And Reagan immediately scrapped any effort of any kind to build out solar power; a nascent industry collapsed almost overnight, with tens of thousands of solar installers laid off. Energy research funding was cut by two-thirds, and it stayed down there for decades. And of course Reagan took down the solar panels from the White House roof; a senior administration official said they were a "joke." They became, in Carter's fateful prediction, a "museum piece"—I've already described seeing them in the collection of Huang Ming, the Chinese solar pioneer. Oil won, and activism lost.

But activism didn't disappear, and indeed it began to blossom again a decade later, after James Hansen shocked the world with his report to the Congress that there was something called "the greenhouse effect." This time it wasn't the US in the lead—the Reagan consensus, and the power of the oil industry, made sure that not much of note happened here. But across the Atlantic in Germany, the Green Party began noisily demanding action on what would eventually be called "feed-in tariffs," which meant generous payments to people producing electricity on their roofs as they returned power to the grid. They started small—the first program, in 1991, was called A Thousand Roofs; by later in the decade the target grew to 100,000, still a fairly modest tally. (Germany has 43 million dwellings.) But the Greens held the balance of power at critical moments, and they had a partner in centrist prime minister Angela Merkel—a former scientist who understood the dangers of climate change. And so over time they built what in German is called the *Energiewende*, or "energy turnaround." In essence, Berlin did on a grand scale what I did in my backyard: paid the early high price of solar power in order to help get it established. It was an act of extraordinary generosity;

the country that had caused so many of the 20th century's problems was doing more than any other to deal with the crises of the 21st. But if the money came from Düsseldorf and Bremen and Nuremberg, the real magic happened in Wuxi and Dezhou: the sudden demand for solar panels was what allowed those Chinese factories to start the cascading cost reductions that got us where we are today. "We created the mass market, and that led to the increased productivity and dramatic decrease in costs," Rafael Fücks, president of the Green Party's political foundation, said in 2015. As Thomas Friedman wrote in *The New York Times* more than a decade ago, that "was a world-saving achievement."

But if it's actually going to save the world—whose temperature has spiked so sharply in that last decade—then it's going to take a lot more activism. There are still two forces inherent in human affairs that haven't gone away: inertia and vested interest. And between them they are entirely capable of slowing this energy transition sufficiently to allow the collapse of the planet's climate system.

That new activism, though, won't look precisely like the environmentalism of the last couple of decades. We have worked hard over those years to oppose the expansion of the fossil fuel industry—to try, for instance, to block approval of the Keystone and the Dakota Access and the Mountain Valley pipelines, Line 3 and Line 5 pipelines, the East African Crude Oil Pipeline, and the Trans Mountain pipeline. (So many pipelines.) And we've tried to slow the permitting of huge new liquefied gas export terminals and to demand that the oil industry stanch the flow of methane from its wells. Some activists have worked hard to put a price on carbon, an important task that proved difficult in tax-averse America. We've also worked hard to raise the cost of capital for the fossil fuel industry—campaigners across the globe persuaded university endowments and pension funds to divest their coal and oil and gas stocks; when Peabody Energy filed for bankruptcy, it cited those campaigns as one of the reasons. We've been arrested outside the banks that offer easy loans to their long-time customers in the hydrocarbon cartel, and we've been arrested trying to expose Big Oil's longtime campaigns of deception and denial, in an effort to tarnish their image and dent their lobbying success. It's been deeply useful work, but it's mostly happened in a world where fossil fuels were cheaper than sun and wind; our job, in part, was to reverse that fact by making oil and gas harder to extract. We were trying to create a vacuum that clean energy could fill, and we were trying to win the ongoing subsidies and research dollars that could drive down the costs of wind and solar.

Now that's happened—not just in Germany, but in the US too, where President Obama put money into solar research (we remember the bankruptcy of panel-maker Solyndra but not the many successes) and then, crucially, when President Biden negotiated and signed the Inflation Reduction Act, which finally put serious cash on the table for clean energy deployment. And now we live in a different world—a world where economic gravity works in our favor, where the economic winds are finally at our back. And that means that activism needs to shift its focus.

Which is easier than it sounds. When you're good at saying no, then yes takes some practice—it's a different muscle. We can't give up on "no," not in a world where giant corporations and their client governments are still doing dangerous things—the radical early days of the second Trump administration required playing defense just to hold on to some of what we'd gained. But the job now is not just to block bad things—it's to build good ones. It's to overcome inertia. Which is hard, especially since the fossil fuel industry is skilled at adding to that inertia. (That's the job of all those pipeline execs that I described helping foment opposition to solar farms across the Midwest or along the Jersey shore.) I'm not going to lay out a detailed plan for activists to follow—that's not how activism works. But I will suggest a few key projects, and perhaps a certain . . . mood. I'll concentrate on the US, which is the terrain I know best. But I think some of these make sense everywhere.

The most basic strategy is simply to get out of the way: give the benefit of the doubt to someone building a solar farm or a wind turbine or a battery. I don't think this advice applies everywhere—on Indigenous land, as I've said, I think the residents have earned the right to look askance (if I was an Indigenous American, given history, I would look askance at *everything*). But the group I belong to—older white people—is the group that's best at suing to stop things, the best at working political connections to slow things down. And sometimes we shouldn't. Decades ago, when I was living in a remote section of New York's Adirondack Mountains (still the place I love best on this earth, and where I spend as much time as I can), some local businesspeople proposed a wind farm on the edge of the Siamese Pond Wilderness. It was a few miles from my house, and a place where I've spent innumerable hours skiing, hiking, loafing; I know those woods as well as anyone. And there was no question that they'd be changed: When you looked up from a beaver pond you'd see blades turning on the horizon; at night there'd be a flashing red light. The concrete base pads would mean chopping down some acres of trees. But never mind—I sat down and wrote an op-ed for *The New York Times* supporting the windmills, on the grounds that the deeper threat to the woods I loved was the disappearance of winter, the ever-hotter Augusts. The main regional environmental group, supported mostly by the donations of vacationers, came out strongly against the turbines (relying on an arsenal of bad faith arguments, including the claim that in a storm the blades would come off and roll miles across hill and dale before they crashed into the local school, which is not how wind turbines work). In the event, they won, and the turbines were not built (yet).

But the episode sharpened my thinking. As far as I could determine, I was working from several principles, things to consider before you hire a lawyer and start suing. The first is, we don't live only in our backyard. We share one, with everyone around the world; America's combustion of fossil fuels floods the homes of Pakistanis and dries the grasslands of Kenya.

And the second is, we don't just live only in our own moment; we're accountable for past behavior. Adirondackers and Vermonters have spent 150 years pouring carbon into the atmosphere at rates that most of the people of the world will never match. So even if I've put a solar panel on my own roof now, I still owe a debt.

The third is, idealism demands a certain amount of realism. Yes, there's always a better place to do something, and always a better way. If we drove half as much, then we would need half as many solar panels. But unless you really have a way to get people to drive half as much, then it's petty to use that as your argument.

Which leads to the fourth and final principle: emergencies demand urgency. If you put up a field of solar panels now, they don't have to be there forever—maybe someday we will develop small, cheap, safe fusion reactors and we can take them down. (And by the way, in every place that I know of solar developers have to put down a bond to pay for that eventual removal.) But if you *don't* put up a field of solar panels now, then the world will break.

These principles won't tell you how to act in every case. But they're hopefully a guide to not being the person who just doesn't want to look at the windmill, and who then builds an arsenal of increasingly disingenuous arguments. I don't want to be that person, not at this stage of our planet's history. And there are plenty of other people who feel the same way. A unique organizing project called Greenlight America has spent the last few years teaching residents how to stand up at public meetings and tell the truth. In Erie County, Pennsylvania, in 2024, they helped defeat an effort to block clean energy development with a widespread lobbying campaign; in western Colorado, in the country around Grand Junction, their Yes to Solar campaign managed to lift a moratorium on solar development. "In the past year we've advanced approximately 4.4 gigawatts of clean energy to the grid through project approvals and passage of favorable local ordinances," its director, Matt Traldi, said. "We're not talking about the US Senate—the nitty gritty work is done in small meetings of county boards. We found that even a relatively small number of people, three people here, five people there, can absolutely change the outcome, and for massive projects that have a hugely outsized impact."

The next job is equally important: We need to keep the money flowing, in order to pay for the one-time start-up cost of renewable energy. In the United States, over the next decade, that means trying to protect the funding in the Inflation Reduction Act. The bill was an unlikely achievement—the Democrats controlled the Senate by a single vote, and that vote belonged to Joe Manchin of West Virginia, who in 2021 had reportedly taken more campaign donations from the oil, coal, and gas industries than any other senator. (Not an easy contest to win.) It looked, for the first two years of the Biden administration, as if Manchin would prevent any deal—and indeed he slowly whittled down an awful lot of the best parts of Biden's original Build Back Better package (itself a very modest version of the original Green New Deal). The legislation went in stages from $3.5 trillion over 10 years to $2.2 trillion to $1.5 trillion, and then, in July 2022, Manchin went on *Fox News Sunday* to say it was all over. "I cannot vote to continue with this piece of legislation. I just can't," Manchin said. "I've tried everything humanly possible. I can't get there." Bret Baier, the Fox News anchor, asked Manchin if this was a definitive no. "This is a no on this legislation," Manchin replied.

But the White House—which had been a supplicant for two years, attending to Manchin with the care usually reserved for a sick toddler—got a bit mean. A few hours after Manchin's interview, the White House press secretary said Biden's words: Manchin's actions "represent a sudden and inexplicable reversal in his position, and a breach of his commitments to the President and the Senator's colleagues in the House and Senate." For Biden, that was strong language. And by all accounts Manchin's Democratic colleagues let him know they were tired of his act too—he'd scuttled the most important legislation the president had proposed.

And so Manchin caved. After secret negotiations with Senate and White House leaders, he suddenly signed on to a new plan—smaller still, and larded with yet more presents for the fossil fuel industry, but nonetheless the first serious response the US Congress had ever made to the climate crisis. I think that, in the end, Manchin understood that his entire legacy would be singlehandedly blocking the most important bill on the most important issue of his time—that his soft and well-manicured fingerprints (the man lives aboard a yacht anchored in the Potomac) would be visible in the geological record of the planet. In that sense, it was a win engineered by everyone who ever wrote a letter to the editor, carried a sign at a march, went to jail blocking a pipeline, voted to divest a university endowment, sent $10 to a climate group, made their book club read a global warming treatise. And it mattered mightily—suddenly the Loan Program Office of the Department of Energy became a whirlwind of energy, sending out loan guarantees to entrepreneurs starting battery factories and building panels and trying to connect new transmission lines. As the end of the Biden administration approached, they were shoveling money out the door as fast as they could, cognizant that Trump's Project 2025 took dead aim at their work. Because of the size of the program—which is designed to last a decade—the broad rebates aimed at consumers took longer to set up, but in many places they were coming online, too, as the Biden years ended. Now the question is whether they'll have broad enough support to survive.

On a global scale we're in an even tighter spot. The world's governments agreed at the global climate talks in 2024 that they would ramp up energy and climate aid to the developing world to $300 billion annually by 2035—that was a quarter what the emerging nations were demanding, and in any event it's entirely unclear if anyone will pay it. (Well, Norway probably will, and maybe Sweden, but their share is pretty small.) It's hard to imagine any US Congress in the next decade spending money on climate aid to poor nations, even though, as I've said, the merest tax on billionaires or on stock deals would provide the necessary funds. And those funds—globally and in the US—are what's needed to make the one-time investment: to put up the solar panels that would then catch free energy. We've subsidized fossil fuels for centuries—the International Monetary Fund said in 2023 that if you added it all up, the world was giving coal and oil and gas $7 trillion a year, above all by letting them use the atmosphere as an open sewer for free. We desperately need to subsidize clean energy for a couple of decades, in order to get ourselves out of the emergency all those hydrocarbons have created.

Money, however, is not the only job here. We also have to reduce the regulations that get in the way of change. Right now, remember, there are more than enough clean energy projects to provide all the power we need just awaiting approval in this country. Some require giant transmission lines that connect the Midwest and Great Basin to the coasts; some require new substations and transformers for local utilities. In every case it's crucial to break deadlocks—activism in these cases means learning how to work through opaque institutions like the Public Utility Commissions that set policy in most states. These have been protected for decades by a force field of sheer dullness: It's hard to get activists together to spend Tuesday afternoons and Thursday mornings sitting in featureless agency offices listening to PowerPoint presentations about interconnection queues, and so industry lobbyists end up controlling these boards. "The biggest problem we have with utility regulation in this country is the capture of independent agencies by utilities," Leah Stokes, who helped write the Inflation Reduction Act (from the neonatal care unit with her premature twins) said; indeed, a study of 473 public utility commissioners across the country found that exactly half went to work for the utilities they'd been regulating. It's a gas-powered revolving door. But we're learning. At Third Act, which I founded a few years ago to bring older people like me into the climate fight, we've discovered that retirees are great at this work—willing to bring the crossword puzzle or crochet basket into hearings, willing to do the work in the trenches.

I confess I find it maddening, even in small doses. In the spring of 2024, I took a trip to the board meeting of the Tennessee Valley Authority as part of an effort to get them to emulate California and Texas instead of building more gas-fired power plants. You would have thought it would be relatively friendly ground—the White House (which at the time meant Biden) appointed most of the members of the board, and they were meeting in Nashville, home ground of Al Gore. But the testimony featured a parade of local officials, each insisting that they needed "reliable" energy from gas instead of those newfangled batteries; they let me talk for two minutes, smiled, and went on to the next. But activism has to mean this kind of patient engagement, just as much as it means going to jail. (Which, trust me, is boring in its own way.)

And when it works, it's intensely satisfying. After endless hearings, the Massachusetts legislature ended 2024 by passing crucial siting and permitting reform laws that will make it easier to build new community solar farms—that's the kind of law that propelled those Illinois developments I described. In New York City, as I was writing these pages, a new congestion pricing law went into effect—I've known the advocates who've spent two decades getting it passed, and within hours it was cutting traffic coming out of the Lincoln Tunnel, dramatically reducing accidents, clearing the air over Lower Manhattan, and raising cash to improve the subways. (Needless to say, the Trump administration immediately tried to kill the experiment.)

The possibilities are endless. In November 2024, for instance, a new study showed that we could double renewable generating capacity in the US without building new transmission lines, simply by siting wind and solar near fossil fuel plants that don't currently use their grid connections all the time. The study findings sound so dull ("also known as surplus interconnection, repowering, cable pooling, or hybridization, it would involve the expedited addition of renewable energy and/or battery storage capacity at or in close proximity to existing power plants"), but overcoming boredom is part of the work of activists. Maybe we could call it "grid recycling"; let's make some gorgeous drawings of solar farms with sheep surrounding the smoke-belching power plant. And this can be done in the reddest terrain. Texas is now leading the country in clean energy buildout in part because it approves new wind and solar farms with a "connect and manage" process: in short, the utility pretty much has to deal with the new power that people want to build. This is one of those times where a kind of cowboy capitalism is actually useful. It's harder on the utility managers who have to scramble to deal with the new power, but it's easier on ratepayers—and, of course, on the climate.

The easiest wins—though these too will be hard-fought—may come at the most local levels, where a clotted regulatory system slows down what should be the simplest chore: putting panels on tops of roofs and batteries in garages and basements. We're terrible at it: A third of homes in Australia have solar panels on top, compared with about 7 percent in America. Remember those 1.5 million Germans who installed "solar balconies" in 2024? The comparable number in the US was about zero. That's not because they're better situated—Florida has a remarkable resource, but it badly trails Germany. The difference is that it costs three times as much to put solar on the roof in America as it does in Europe or Australia, and that's mostly down to what the industry calls soft costs, which is to say everything that isn't a panel or a wire.

I sat down not long ago to talk about all this with an old friend, Billy Parish, who can lay a decent claim to being the original American climate activist. He dropped out of Yale in 2002 to form the Energy Action Coalition, which organized students at college campuses around global warming; eventually he left that work to others and founded a company called Solar Mosaic that is essentially a lending platform for solar projects. Beginning with a few churches and then an Oakland food bank (Prince helped fund that one; his drummer, Sheila E., was from the area), Parish has grown it into a large-scale business, which has funded close to 10 percent of all the residential solar in America. The model's pretty simple: They connect with contractors around the country, and design systems that save money for their clients. When all is said and done, perhaps the homeowner pays $200 a month on the loan from Mosaic, which might represent a $50 savings from what they were previously paying for electricity.

    The problem is, the price for putting solar on rooftops remains high enough that the savings are pretty marginal in large parts of the country. And the costs are high because it takes forever to get the things permitted. In any given town, a building inspector has to come out, maybe several times. A huge percentage of potential customers drop out because it's takes so long—customers where, as Parish says, "you've already done multiple visits, rolling trucks to the site." And the cost of capital adds up over those months too—altogether, says Parish, permitting adds about $7,500 in wasted cost to the average solar system. Wasted because these systems are fundamentally safe—as a onetime volunteer fireman, I can tell you that they are the least of our worries. (Talk to me about creosote buildup in chimneys with woodstoves. Then talk to me about space heaters.) In Germany (which is not exactly known for its loosey-goosey attitude toward rules and regulations), the law now lets you put in solar systems up to a certain size with no permits at all; in Australia, installers "self-permit" their systems, and a government official comes along to audit one in 10 to make sure they're not cutting corners.

    In America, the easy option is something called the Solar Automated Permitting Process (SolarApp), a piece of software developed by the federal renewable research lab and provided free to communities that lets them do quick approvals of solar panels, of batteries, of heat pumps. Nick Josefowitz, who's head-ing a national effort to ease permitting, said that when the new software went into use in Tucson, the rate of rooftop solar installations doubled. California and Maryland have now passed laws mandating its use, but that leaves 48 states to go—and many of them are the kind of places where a little rhetoric about Big Brother should go a long way. I mean, we let you own a semiautomatic rifle with no questions asked, and you're free to drive a giant Escalade down the street even if you can barely see over the front bumper. But you can't hang a solar panel from the balcony of your apartment? Don't tread on my arrays, dude.

"I feel like in America the government has stopped trusting people to do the stuff they do really well," says Josefowitz. "If you're a solar installer and you've installed hundreds of these things, the government treats you like you're an incompetent crook trying to hoodwink people into putting a fireplace on their roof." It all adds up: "If General Electric had to produce a slightly different washing machine in 18,000 different cities and counties, washing machines would cost $20,000," Josefowitz said. He'd recently taken a trip to the European headquarters in Brussels, which is sometimes seen as a nest of bureaucrats. "But when I sat down with the EU staffers in charge of their distributed energy policy, my questions were completely lost in translation. 'You're trying to streamline solar rooftop permitting? Why would you have any rules on this? We're trying to streamline for one megawatt projects—that's like four Amazon warehouses.'"

In some ways, this should be the easiest kind of activism, because it's pushing on an open door. People dislike regulation. And around the world—despite the best efforts of people like Trump—people intuitively love clean energy. A global poll in 2023 found 68 percent of us favoring solar energy, "five times more than public support for fossil fuels." Even in the US, 2022 surveys found almost 70 percent of adults backing more sun and wind over more fossil fuel—that was down from 80 percent two years earlier, reflecting the endless Republican propaganda campaign. But those kind of numbers mean that the will is there; we just to need to make the process affordable and easy. And there are encouraging signs—in March 2025 the deep-red Utah legislature voted (unanimously!) to make the beehive state the first in the country that won't require permits or fees for balcony solar.

Part of that is simply helping each other. Rewiring America, a nonprofit set up to make it easier for households to access the funding to install things like heat pumps and batteries has the usual savings calculators and guidebooks online, but they've also set up classes for "navigators." In the words of executive director Ari Matusiak, these navigators are trained to serve as "kind of like the right hand to someone and help them through the process." And what's amazing, he went on, is that it "taps into a desire from so many people in supporting this transition. We started off with this electric coaches program. It's like nine hours of classes that I went through and I barely passed. But we've had almost a thousand people just raise their hands and say they want to be part of it." Another nonprofit, Solar United Neighbors, has helped whole blocks come together to put up panels all at once, saving serious money; they've also set up a help desk offering "solar expertise, agenda-free." As organizer Jani Hale explained, "If we don't already have the answers we start crowdsourcing with our in-house solar experts, some of whom have the same equipment, panels, inverters, EVs."

Activists, in my experience, are extremely bad at one thing, and that's celebrating. Their hesitation is understandable, because there is always more to do, and organizers tend to think: "If we say we've accomplished something, it will take the pressure off." But my sense is that people are more eager to join winning teams. That's one reason that even as I'm writing these pages in early 2025 I'm working with many others to launch Sun Day, which should be taking place just about the time this book is published—on the weekend of the autumnal equinox. It's an effort to re-create some of the passion and joy (and anger) that marked the first Earth Day, way back in 1970, when 20 million Americans took to the streets. It won't be that big, of course—but we hope that lots of the solar-powered homes in this country will be displaying green lights that day, and that we have long parades of e-bikes, and that we're throwing the switches on plenty of new community solar farms and wind parks. We need cities sending their electric buses out decked in ribbons, and we need concerts and sunflowers and prayer services and teach-ins and people climbing iconic buildings to do guerilla installations of solar panels. We'll be pushing for precisely the kind of things I've been describing in this chapter—relaxed regulations, more money, quicker action. A portion of my earnings from this book are going to support our organizing, so in some sense you're already helping out.

This Sun Day effort seems to be finding enthusiasts elsewhere around the world, too—though everywhere the demands are different. In Africa and South America people need different things than they do in Germany and Vietnam, but everywhere the vision converges. And that vision, I think, is not of "net zero by 2050," or "dramatic reductions in carbon emissions," or any of the other phrases that have come to define the climate debate. It's not that those things aren't important—they are. It's that they don't really offer a positive vision of the world we might build, the kind of vision that could motivate a much broader swath of people. I did my best to define that vision earlier in this book: a world where we no longer set things on fire, but rely instead on the great fire out there in space. A world where we turn to our star. I find that vision deeply resonant: I think it connects with the oldest, strongest currents in human history.

And I think, as I'll also describe, that this switch carries huge liberating potential—that it could help build the kind of communities we most want. It offers as well the chance for a group project, something to work on together in this incredibly divided country and world. For Americans, the last project like that was the moon shot in the 1960s—but really, most of us just paid our taxes and watched the engineers and astronauts do their thing. It made us proud, by and large, and it yielded big benefits; in retrospect, the most important weren't Teflon or Tang, but that photograph of the blue-white earth spinning in the black void of space. This project is far bigger, involving every one of us on this living marble. It's a fight for survival, but also for something literally larger than ourselves. If there is one thing we have in common, it's that we live beneath the same sun; learning to face it together, instead of tending the myriad individual flames we've lit over the last few centuries, might connect us in ways that we dearly need.

But before we focus on all that, I want to describe the one truly useful kind of large-scale fire, the flame (besides barbecues and campfires and candles and Olympic torches) that we'll still need. That's the use of flame to control flame and to manage land—a skill developed over many millennia by the original inhabitants of much of the world.

Of all the fires burning on earth, none are more terrifying than the conflagrations that light the arid West, the Mediterranean, the eucalyptus forests of Australia, the boreal woods of Siberia and the Canadian north, and the familiar streets of LA. None of it is surprising, of course. Los Angeles had 0.29 inches of rain and some of the hottest temperatures ever recorded in the eight months before the 2025 fires; how could it not be a tinderbox? California, thanks to Hollywood, was once literally the planet's idyll, the place where people imagined the easy life must reside; now, as the state's governor told reporters the morning the fire broke out, "We don't have a fire season anymore. We have a 'fire year.'" But it's not just the West. I'll always remember waiting to be arrested outside the White House at a sit-in to protest a new pipeline in the summer of 2023 as DC experienced the worst air quality day in its history—wildfire smoke pouring down from record Canadian fires meant that we could barely see a hundred yards across the lawn to the Oval Office. In Siberia, even winter's icy cold is not enough to blot out the blazes; researchers reported "zombie fires" smoking and smoldering beneath feet of snow. There's no question that the climate crisis is driving these great blazes—and also being driven by them, since they put huge clouds of carbon into the air.

There's also little question that in many places the fires, though sparked by our new climate, feed on an accumulation of fuel left there by a century of a strict policy which treated any fire as a threat to be extinguished immediately. That policy ignored millennia of Indigenous experience using fire as a tool, an experience now suddenly in great demand. Indigenous people around the world have been at the forefront of the climate movement, and they have often been skilled early adopters of renewable energy. But they have also, in the past, been able to use fire to fight fire: to burn when the risk is low, in an effort to manage landscapes for safety and for productivity.

Frank Lake, a descendant of the Karuk Tribe indigenous to what is now Northern California, works as a research ecologist at the US Forest Service, and he is helping to recover this old and useful technology. He described a controlled burn in the autumn of 2015 near his house on the Klamath River. "I have legacy acorn trees on my property," he said—meaning the great oaks that provided food for tribal people in ages past—but those trees were hemmed in by fast-growing shrubs. "So we had twenty-something fire personnel there that day, and they had their equipment, and they laid hose. And I gave the operational briefing. I said, 'We're going to be burning today to reduce hazardous fuels. And also so we can gather acorns more easily, without the undergrowth, and the pests attacking the trees.' My wife was there and my five-year-old son and my three-year-old daughter. And I lit a branch from a lightning-struck sugar pine—it conveys its medicine from the lightning—and with that I lit everyone's drip torches, and then they went to work burning. My son got to walk hand-in-hand down the fire line with the burn boss."

Lake's work at the Forest Service involves helping tribes burn again. It's not always easy; some have been so decimated by the colonial experience that they've lost their traditions. "Maybe they have two or three generations that haven't been allowed to burn," he said. There are important pockets of residual knowledge, often among elders, but they can be reluctant to share that knowledge with others, Lake told me, "fearful that it will be co-opted and that they'll be kept out of the leadership and decision-making." But, for half a decade, the Indigenous Peoples Burning Network—organized by various tribes, the Nature Conservancy, and government agencies, including the Forest Service—has slowly been expanding across the country. There are outposts in Oregon, Minnesota, New Mexico, and in other parts of the world. Lake has traveled to Australia to learn from aboriginal practitioners. "It's family-based burning. The kids get a Bic lighter and burn a little patch of eucalyptus, the teen-agers a bigger area, adults much bigger swaths. I just saw it all unfold." As that knowledge and confidence is recovered, it's possible to imagine a world in which we've turned off most of the man-made fires, and Indigenous people teach the rest of us to use fire as the important force it was when we first discovered it.

Amy Cardinal Christianson, who works for the Canadian equivalent of the Forest Service, is a member of the Métis Nation. Her family kept trapping lines near Fort McMurray, in northern Alberta, but left them for the city because the development of the vast tar sands complex overwhelmed the landscape. (That's the 173 billion barrels that Justin Trudeau says no country would leave in the ground—a pool of carbon so deep that the climate scientist James Hansen said pumping it from the ground would mean "game over for the climate.") The industrial fires the oil from those tar sands helped stoke in turn helped heat the earth, and one result was a truly terrifying forest fire that overtook Fort McMurray in 2016, after a stretch of unseasonably high temperatures. The blaze forced the evacuation of 88 thousand people, and became the costliest disaster in Canadian history.

"What we're seeing now is bad fire," Christianson said. "When we talk about returning fire to the landscape, we're talking about good fire. I heard an elder describe it once as fire you could walk next to, fire of a low intensity." Fire that builds a mosaic of landscapes that, in turn, act as natural firebreaks against devastating blazes; fire that opens meadows where wildlife can flourish. "Fire is a kind of medicine for the land. And it lets you carry out your culture—like, why you are in the world, basically."

9

A Subtly New World

I'm not old enough to have known Woody Guthrie, but I've been to his museum in Tulsa and seen his guitar, with the phrase "This Machine Kills Fascists" written above the sound hole. I did know (and love) Guthrie's great compatriot Pete Seeger, and I've watched him play a banjo emblazoned with the slogan "This Machine Surrounds Hate and Forces It to Surrender."

I'm not going to make claims quite that grand for these new machines, the solar arrays and slowly spinning wind turbines, the batteries and heat pumps and EVs and e-bikes. But I want to suggest that they're probably the most potentially liberatory technology loose in our world, the single thing most likely to subvert the cartoonish inequality and eat away at the grotesque privilege that dominates our world. (It's worth noting that Donald Trump banned new wind projects the same day that Elon Musk offered what for all the world looked like a Nazi salute.) Solar panels and wind turbines won't liberate us automatically, or by themselves, but they allow us to at least imagine a new trajectory, away from the domination that seems to be closing in on all sides.

At heart that's because, as I've intimated before, energy from above is fundamentally different from energy from below. Coal and oil and gas are found in deposits, closely related to the forests and swamps of the Carboniferous; these are not evenly distributed around the world, and as I've also pointed out 80 percent of humans live in countries that must import fossil fuels. Even within the exporting countries—the United States, Russia, Saudi Arabia—the deposits are geographically concentrated, and the riches and power they produce tend to flow to a small cohort. As Samuel Miller McDonald pointed out in his 2024 book *Progress*, as the 19th century began, the richest 1 percent held just 8.5 percent of America's wealth; by the time it ended, the top dogs had 50 percent of the money, "partly thanks to fossil fuels, which could be easily concentrated, controlled, and transformed into liquid capital by a small management class." (Meet John D. Rockefeller.) This trend has continued to the present, and around the globe. As McDonald points out, "Because fossil fuels themselves are easy to concentrate, they often yield authoritarian outcomes." A petrostate, he shows, is "fifty percent more likely to be authoritarian and only a quarter as likely to transition to democratic government than a state without petroleum as a major economic base." (Meet Vladimir Putin, meet Mohammed Bin Salman and his bone saw, meet the Koch brothers.) Just the way we talk about coal and oil and gas makes the point: They are hoarded in reserves, like money in a vault.

Whereas, again as I've said before, sun and wind are by definition universal (and the places with the least sun often have the most wind). Instead of being concentrated they are diffuse—sunlight falls everywhere, and while some places are sunnier than others there is enough everywhere to get the job done. That's why Germany and the Netherlands vie with Australia for the most solar PV per capita, and why Massachusetts trails just Hawaii in the same category. (Rhode Island is ahead of Arizona.) That diffusion used to be the weakness of the power from the sun—a barrel of oil will always have more immediate energy than a meter of sunny grass. But if you have doubts about the wisdom of our current concentrations of political and economic might, then diffuseness is a feature, not a bug. It's the difference between, say, a centralized information system and the internet; just as information now flows from all directions, so can electrons. (And the advantage here is that we don't have sexist or cruel or racist or idiotic electrons; electrons don't dox and they don't know from GIFs). In fact, these new energy sources seem to me to hit a happy medium, not so concentrated that they build up centers of unshakable power, not so spread out that they take up the entire landscape. The factories capable of producing solar panels are complex and for the moment concentrated in China, but it's not rocket science either—as India and the US are beginning to demonstrate, it's entirely possible to match the Chinese at this task. And once you have the panels, the control begins to disappear: There are hundreds of thousands of small contractors competent to put them up. Relatively small firms can build relatively small solar farms; the capital required is an order of magnitude smaller than drilling a new oil field or building a refinery or a pipeline system.

There will be plenty of solar and wind millionaires and billionaires—there already are. But the chance for Rockefeller wealth is slim. (Elon Musk may be the weird exception to this rule, though people around the world seem to be finding other EVs and batteries since he turned at least temporarily Trumpish.) But for every Musk there are, in places like Germany and Scandinavia, a thousand churches and labor unions and community groups that control wind turbines or solar farms. We could do far more to make sure that these new technologies are publicly owned. A 2024 report from the Transnational Institute looked at case studies from Uruguay to the UK and concluded that the renewable energy transition is "an opportunity to remake the electricity sector so that it prioritizes green energy, affordability, and equity—and an empowered, democratic public system could deliver that future" best. So it doesn't solve our inequality problem overnight, but it seems to me that it puts us on a new trajectory.

And that new trajectory, once it gets established, is harder to shake. When Vladimir Putin invaded Ukraine, energy was his great weapon: He had Europe over a literal barrel. But Europe responded by dramatically accelerating its green energy transition—that's why Germany, to give one small example, radically relaxed its Teutonic rules about balcony solar panels. And in Ukraine itself, there's been an interesting test of these new technologies. The man who owns the country's largest private energy company, Maxim Timchenko, told the Associated Press in November 2024 that when the Russians target one of his solar farms with their missiles it takes about seven days to swap out the broken panels and get fully back online; a similarly scaled attack on one of his gas-fired facilities takes four months to fix. "That's the difference between centralized and so-called decentralized generation," Timchenko explained. "It's much more resistant and difficult to destroy"—and even when you knock out some panels the rest go on working. (Ukraine's biggest nuclear reactors at Zaporizhia of course shut down after the invasion, which is a good thing given the tank battles that have been fought in the vicinity.) My dear colleague Svitlana Romanko, who used to work for the Catholic climate movement, spent the war years creating a remarkable nonprofit called Razom We Stand; it has a plan for rebuilding Ukraine with clean energy after the war, and indeed in the fall of 2024 it brought a dozen Ukrainian mayors to Washington and Wall Street in search of investors. Their goal, says Romanko, is that Ukraine becomes "the first post-war country in the world to have renewable energy as the basis of its reconstruction."

As much as I like the solar panels on my roof, I'm even more interested in the ways all these new possibilities connect—the ways they might build communities both intentional and ad hoc. We're used to the idea that things either belong to us (our car) or that they belong to everyone (a subway car) but there's a third category—things that are ours, but that we let others use, often for a price. Think Airbnb. A far more important (and hopefully less abused) example of this third category is going to turn out to be things like the batteries in our cars and homes, and even our clothes dryers. We own them, but we only use them some of the time—which is a waste. And waste creates opportunity, in this case "virtual power plants."

On any given Monday here in Vermont, for instance, Josh Castonguay, the vice president of innovation at our utility, Green Mountain Power, studies the forecast for the days ahead, asking questions like, "What's it looking like from a temperature standpoint, a potential-of-load standpoint? Is there an extremely hot, humid stretch of a few days coming? A really cold February night?" If there is trouble ahead, Castonguay prepares Vermont's single largest power plant, which isn't exactly a power plant at all—or, at least, not as we normally think of one. It's an online network, organized by the utility, of 4,500 storage batteries (currently, most of them are Tesla Powerwalls), spread out across more than 3,000 Vermont homes. The network also includes a broad array of residential rooftop solar panels, which produce much of the energy stored in those batteries, and smart water heaters and EV chargers. The people who have these assets aren't off the grid; they're Green Mountain Power customers who, for a discount on their bills, agree to plug their batteries and appliances into the utility's network and to let the company control the devices so that they use less power at critical moments. (If customers need to override the company's commands, they can.) This means that Castonguay (or, really, his algorithms) can program storage batteries to be charged 100 percent before a storm hits. Or, if it's going to be a hot day, he can preheat water heaters in many homes in the morning, so that in the afternoon, as the temperature rises, more power will be available to run air conditioners. He can also precool some big buildings in the morning. "Then, if you think about it, the building itself is the battery," he said, in the sense that it stores chilled air for later in the day. "We have about fifty megawatts" of this distributed power, Castonguay told me. "At the scale of Vermont, that's a lot."

Utilities have always been able to dispatch supply—when demand rises they bring power plants, which are often in idle mode, online as demand requires. Now they're increasingly able to dispatch demand as well, turning down thermostats or delaying car charging. And it's not just in Bernie-Sanders-socialist-Vermont where this is happening: In November 2024 Google Nest announced plans for what will be the biggest virtual power plant yet, providing a gigawatt's worth of power in Texas. Customers will get a free smart thermostat and sign up for a bargain payment plan that allows the company to shift their usage just a tad during periods of stress—a "smart plug" can even turn a fridge into a big help. Each home might contribute an almost unnoticeable kilowatt or so, but multiplied by millions of homes that's a lot of power.

I started writing about all of this back in 2015; in those days Mary Powell was the CEO of Green Mountain Power, and setting up some of the first pilot projects in the country. "You wouldn't notice that we're turning down the water heater for just a few seconds," she said. But getting permission from customers requires "a different kind of relationship. Can we really build a deep emotional and intellectual relationship with our customers?" She's gone on to head Sunrun, the nation's biggest supplier of rooftop solar and increasingly of basement batteries; already, she says, the company is "sitting on top of 7.6 gigawatts of solar and 1.8 gigawatt-hours of storage." Their customers are now auto-enrolled as power suppliers, she says, and almost no one opts out—in part because they get a check for $700, and perhaps in part because they understand they're helping solve a real problem. Jigar Shah, the dynamic entrepreneur that Biden named to hand out the money from the Inflation Reduction Act, says, "If we could fluctuate demand as we do supply" across the country, Americans would save 20 percent on their power bills. Put another way, the Rocky Mountain Institute estimates that by 2030 virtual power plants could reduce peak demand in the US by enough to power every home in California, Texas, and Florida combined. If all the cars in the US were EVs, their batteries hooked together could power the country for two days. We just need to be a tiny bit less individualistic, just bend a bit in the direction of community. The battery on my Kia can cool my neighbor's beer before the Red Sox game.

This idea of doing things together is the kind of solar liberation that appeals most to *me*—I mean, I'm a sometime Sunday school teacher. For the more technically inclined, though, the prospect of essentially unlimited amounts of all-but-free energy opens the mind in other ways—as *The Economist* put it, "In its radical abundance cheaper energy will free the imagination, setting tiny Ferris wheels of the mind spinning with excitement and new possibilities." If this scares you a bit, it probably should, given that our recent big energy-consuming projects are things like Bitcoin or ChatGPT, the latter expensively churning out prose so limp it darkens the heart. But there actually are remarkable possibilities. *The Economist* again: "One way to drastically reduce the spread of airborne disease is to speed up the rate at which the air in the world's buildings is vented and refreshed. If energy is expensive, this is not feasible. But what if?"

In parts of the world that have begun to seriously build out solar and wind power, we're already seeing parts of every day when there's such an abundance of clean electrons that their price drops to almost nothing. (Sometimes literally nothing; sometimes, due to the oddities of the utility billing system, less than nothing.) And so we're beginning to get a few clues as to what that world might someday look like.

One day in the summer of 2024 I was in California with an old friend named Danny Kennedy, a veteran guru of renewable energy currently running a nonprofit called New Energy Nexus which helps start-up firms. He'd been pestering me to visit a new company he liked in Oakland, and so we found ourselves in that city's old industrial zone, on a street of battered warehouses.

Kennedy opened an anonymous door and beckoned me inside, into a workshop that seemed like something out of *Gulliver's Travels*. As we came in, three people in lab coats and welder's glasses shouted out a caution: "We're pouring molten metal over here," one said, and indeed there was an orange glow coming from the small crucible she held. Before long the two proprietors of this small factory found us—Alex Grant and Jacob Brown, one a battery expert and the other a savant of lithium brines. They'd once worked at Tesla, but now they had this new enterprise—Magrathea, named of course for the planet that builds other planets in *Hitchhiker's Guide to the Galaxy*. "We see our role as rebuilding the planet after humans almost destroyed it," Grant explained. Rebuilding it out of . . . magnesium.

Magnesium is, it turns out, the world's third most important structural metal, behind steel and aluminum. Well behind—but if you could produce it cheaply there'd definitely be a larger market; Grant curated a small museum of magnesium products on some shelves in the cluttered workshop and it featured snowshoes, a bike, a lawn mower. And the dashboard of a Corvette. "Magnesium and plastic inhabit the same density regime," he said. "It's a third lighter than aluminum, four times lighter than steel. We've had global automakers tell us, 'If we had a supply chain, it'd be a no-brainer. It's lighter, easier to die-cast, stronger. All the fundamentals point to it being at least the number two metal." But right now it takes a lot of energy to make—China produces most of it, in smelters they have to heat to 1,100 degrees. The process Magrathea uses requires 400 degrees less heat—and it exploits one of the unique characteristics of magnesium. Unlike temperamental aluminum (once you start smelting it you can't stop, because if the temperature drops even a little, it crystallizes), magnesium is hardy. "We have 100 degrees of delta before magnesium freezes," Brown said, which means that you can run your smelter when the sun is providing cheap energy and then just turn it off when the sun goes down. We were standing in front of a squat kiln with a thick electric cord going in, where the process was underway. But it was coming on to early evening now, so it was time to turn it off. They'd switch it back on again when the morning sun rose high enough that the price of electricity dropped toward zero.

"People are used to a continuous electricity supply," Brown said, handing me a can of LaCroix water (which he noted contained magnesium in both the water and can). "We're designing processes to make use of intermittency. Intermittency makes magnesium the metal of the future." As they scale up, he said, they'll almost certainly build their first big plant somewhere in the Midwest with excess wind power. 'We'll just turn off when people come home and turn on their washing machines."

And here's the kicker. You don't actually need to mine *anything* to make this process work. Not coal to provide the heat, and not ore to provide the magnesium. That's because magnesium is the second-most common mineral in saltwater (after the sodium that's now being used for batteries). "You don't need to knock down Indigenous sacred sites in Western Australia or cut down rainforests in the Amazon to mine," Grant said. One hundred and forty two gallons of seawater gives you a pound of magnesium.

"That means you can use this process anywhere—on the African shore, in Australia," Danny Kennedy said. Anywhere there's seawater and sunlight, because that's literally all it takes.

10

Face to the Sun

This book, I think, has been impeccably rational so far—all kilowatt hours and insurance premiums, all geopolitics and climate science. And it might make sense to simply end here, having proved, at least to my satisfaction, that our civilization would be well-advised to deploy energy from the sun just as fast it can. But a feeling has been growing in me as I worked on gathering the material for this book in these last years—a sense that just as the sun and wind can be economically and politically and environmentally liberating, it also might have a . . . spiritual dimension. I use that word reluctantly—*emotional* or *psychological* are probably less laden ways to say the same thing. But here's what I mean.

When, in my 20s, I was writing *The End of Nature*, it was mostly an attempt to introduce readers to this new science about what we then called the greenhouse effect. But it was also a lament for the world we were creating—a world where nothing would any longer be "wild" in the old sense, because humans were changing every square meter of the earth with our emissions, making its plants and animals, its seasons and moods reflect our appetites and economies. I thought that this would erode our already weakened connection to the natural world. "I love winter best," I wrote, "but I try not to love it too much, for fear of the January perhaps not so far distant when the snow will fall as warm rain. There is no future in loving nature." I was right in the specific—as I said at the outset, the very day I began to write this book featured a warm winter rain—but I fear I was also right in the general. We look up ever more rarely from the phones in our palms to see the world around us. As *Outside* magazine reported not long ago, we're spending ever less time outside. Since 2008, the annual total of American hikes and excursions dropped by about a billion a year. We're not spending more time with each other (that's dropped even more sharply). We're just wrapped up in whatever we call the hybrid of ourselves and our screens.

So I want to suggest—just suggest—that turning decisively to the sun for the energy we need might come with one more benefit. It might begin to reconnect us with the single most powerful and charismatic object in the natural world, that glowing sphere that rises in the east and transits across the sky each day. Perhaps the sun would at least begin to *register* against those glowing screens that mostly absorb our attention. And that might be a step in the right direction.

Nietzsche, in one of the defining passages of modern literature, declared that God was dead. But it's worth reading that passage to see the language he uses. A madman is running through the market, shouting "I seek God," and the people are laughing at him. "The madman jumped into their midst and pierced them with his eyes. 'Whither is God?' he cried. 'I will tell you. We have killed him—you and I. All of us are his murderers. But how did we do this? How could we drink up the sea? Who gave us the sponge to wipe away the entire horizon? *What were we doing when we unchained the earth from its sun?*"

Nietzsche was writing for the Western world—not all the planet's societies are as deeply alienated, though I must say that when I talk with Indigenous elders and village chiefs even in remote and far-flung places I find them worrying about this same thing. (Find me a corner of the earth without a smartphone.) And Nietzsche was not writing about fossil fuel—but he was writing about a century into the fossil fuel era, when Watt's machines had begun to fundamentally disconnect us from the old world that ran on sun and wind. That disconnection has only deepened and grown more dangerous—what, after all, is the climate crisis but a warped and degraded relationship between our species and the sun, whose rays we are currently trapping in our blanket of greenhouse gases?

As is often the case, the way to deal with a screwed-up relationship is not to break it off—it's to deepen it. We didn't actually turn off our connection to the sun, of course; like all the other animals we're still tied to it in ways we barely comprehend. Seasonal affective disorder is very real; 40 percent of Americans feel their mood slip as the sun wanes. (One hundred percent of humans feel joy when the sun breaks through the clouds.) It was Carl Jung, 40 years after Nietzsche, who sat with a Pueblo elder and asked him what he thought of the sun. "The sun is God," the man said. "Everyone can see that." A few months later Jung was in the mountains above Nairobi, watching for dawn with the locals who waited each day for "the birth of the sun in the morning." As one man told him, "The moment when the sun appears is God." If anything is primordial, it's the sun that hovers above us all. So grant me a little latitude to explore here. It's not crucial, unless it is.

We don't have to look very hard to find the traces of sun worship in human history. (On the day those Bell Labs investigators unveiled the first solar cell, *The New York Times* didn't get three paragraphs into the story without a reference to "this modern version of Apollo's chariot.") I'm just going to randomly list what a few months of nighttime reading has left me with from various Indigenous and folk traditions around the world (traditions, by the way, that seem more sensible than many features of our current Western world). I'm aware that a slapdash list of this sort is the despair of the anthropologist, and that it would be far more respectful to treat each story in depth—but that's not my task here. My task is to say: Everyone before us thought long and hard about the sun. Richard Cohen, in his comprehensive book *Chasing the Sun*, described tribes of Peru and northern Chile who knew the sun as the god Inti, "who descended into the ocean every evening, swam back to the east, then reappeared, refreshed by his bath." Huitzilopochtli, the Aztec god of the sun, is a hummingbird, albeit one in full battle dress, constantly fighting off the darkness; his brothers are the stars and his sister the moon. The Mayans consider the sun a rattlesnake; Hopi creation stories center on the solar god Tawa, often depicted in a round headdress; a raven brought the sun in a bag in some Inuit accounts; in the Yuwaalaraay tradition in Australia the sun was created by throwing an emu's egg into the sky where it burst into flame. The |Xam Bushmen of central South Africa hold that there was once a man whose shining head illuminated the world, but he was lazy and took to sleeping late; his head was cut off and tossed into the sky. Friedrich Max Müller, the 19th-century German scholar who produced a 50-volume English translation of the "sacred books of the East" (and according to Cohen may have been the model for George Eliot's pedantic Dr. Casaubon in *Middlemarch*) thought the sun was the basis of most systems of mythology—after all, one had to explain its disappearance each evening and its arrival again in the morning. In some places a hippo had eaten it, in others a crocodile; in China, eclipses were a dragon devouring the sun, or a hungry dog called Tiangou, who might be scared by banging drums; Fenrir, the Norse wolf, will swallow the sun in the final battle of good and evil.

Chinese altars that date back 5,000 years were aligned with the points where the sun rose and set on the solstices; the tower of the main temple at Angkor Wat aligns with the rising sun of the spring equinox. At midsummer at Stonehenge the sun rises over the Heelstone and shines into the center of the stones; at the Mayan temple at Dzibilchaltun the rising sun at the equinox illuminates the door of the Temple of the Seven Dolls; Hovenweep Castle, on the Colorado-Utah border was apparently used by the ancestral Puebloans to track the peregrinations of the sun across the seasons. The rays of the rising sun at the summer solstice pour through the window of Machu Picchu's Temple of the Sun. (New York City's cross streets are aligned such that the sun sets perfectly on the centerlines every May 28 and July 12, but that's just happy accident—Manhattanhenge. Torontohenge is February 16, and Chicagohenge, March 20.)

If all of this seems deep in the past, that's not entirely true. I have friends who are Sun Dancers in the various traditions of the North American Plains Indians; for some, it's been a key part of their reconnection with their Indigenous roots. Still, there's little doubt that the central role of the sun has faded in most places over the millennia. Look to the Middle East, for example, the place we sometimes think of as the cradle of civilization. The sun was pre-eminent for the Sumerian civilization; Utu, the sun god, and his Akkadian successor Shamash could see, as they traversed the sky, all that happened on earth, making them the arbiters of justice and equity (Shamash gave King Hammurabi of Babylon the famous law code that served as the model for so much that came after). Sun worship may have reached its apogee next door in Egypt, where Ra, the king of the gods, sailed the sky in his solar boat with his sun disk on his head. Pharaohs were considered his sons; the great temples at Karnak and Luxor captured the sunrise at the solstices, and many of the tombs in the Valley of the Kings were aligned toward the setting sun. The shape of the pyramids themselves is likely symbolic of the rays of the sun as they spread down from the heavens.

This region no longer belongs to these gods, of course. Instead it's firmly in the grip of the various religions of the book—Judaism, Christianity, Islam—which frown on anything even resembling sun worship. And yet, as Freud argues in *Moses and Monotheism*, there's a sense in which "modern" religion comes in a direct line from that earlier era—that monotheism really originated in Egypt and that Judaism just dropped the sun. Moses, he insisted, was not Hebrew but an Egyptian follower of a monotheism based on the sun, "who chose the Jewish people to keep alive an advanced ethical and religious belief which the Egyptians were abandoning." Scholars don't accept this version as historical truth, but those of us who grew up with the Bible understand Freud's deeper reasoning—the sense of light as a metaphor for the divine, the idea that God, like the sun, is a source of life and energy. Yes, the Bible is harsh on idolators: "Beware lest you lift up your eyes to heaven, and when you see the sun and the moon and the stars, all the host of heaven, you be drawn away and worship them and serve them." And yet "Sun of Righteousness" is the name that the Hebrew prophets used to forecast the Messiah. Though I'm not a preacher, I lead the Christmas Eve services in our tiny Vermont church, and I always make sure we sing "Hark! The Herald Angels Sing," in part for those lyrics:

> Hail the Sun of Righteousness!
> Light and life to all He brings,
> Risen with healing in His wings.

Christmas, in fact, is the reminder that none of our meaningful traditions are ever fully disconnected from the past. Was Jesus born on the 25th of December? The Bible doesn't say, and early accounts—Clement of Alexandria, say—suggest April or May. But there was an existing Roman solstice feast (Saturnalia) and as Constantine was baptized into the new religion, customs began to be absorbed. (Cultural appropriation!) In 354, Liberius, bishop of Rome, picked December 25 to commemorate Christ's birth, and for many centuries Christmas continued pretty much as the drunken feast around the darkest days of the year that it always had been (even in the New World the Puritans frowned on Christmas, which they called Foolstide). It's taken 2,000 years to wring most of the old sunlight out of the day.

But hey, what do you know, the Roman Sun Day became the Christian day of worship. There's no escaping it. I'm a Christian—a Methodist, the least excitable of all denominations. That means I worship the Son, not the sun: The radical code that Jesus laid out, with its insistence above all on caring for the poor and vulnerable, works for me. But my form of the faith (increasingly remnant in modern America where a cultish and brutal Christianity is now the norm) is perfectly compatible with some low-key reverence for the sun. I am reminded constantly of Francis of Assisi and his "Canticle of the Sun."

> Praised be You, my Lord, with all your creatures;
> especially Brother Sun, who is the day, and through whom
>   You give us light.
> And he is beautiful and radiant with great splendor,
> and bears a likeness to You, Most High One.

It's worth noting, in fact, that the recently deceased bishop of Rome took his papal name from that earlier Francis. And when he wrote his radical 2015 encyclical on climate change, *Laudato Si'*, he took that phrase ("Praised be You") straight from Francis of Assisi's canticle (and that in 2024 he committed the Vatican to become the first state on earth to run entirely on solar power).

As the author of Ecclesiastes (Solomon, at least by legend) put it:

What has been will be again
What has been done will be done again.
There is nothing new under the sun.

As with faith, so with art. From its earliest moments the sun has mattered: We think of the caves at Lascaux, with their magnificent thundering herds of painted bison, as dark and torchlit—but they were also aligned with the solstice sun. Light is the medium of art; it's no wonder that the sun has always fascinated artists. The great scientist/artist Leonardo da Vinci, in his notebooks, wrote "*Il sole no si muove*" ("The sun does not move") 40 years before Copernicus made the same point (Leonardo also invented an early solar hot water heater). But it was left, among the moderns, for Van Gogh to truly channel our star—he'd gone to France from Holland in 1888 to seek a "stronger sun," and he found it from the house at Arles where he painted first the surrounding sunflowers, and then in the two canvases of seed sowers at sunset he managed somehow to depict the sun as it never had been captured before—low in the sky, a source not of power so much as of joy. As the German critic Wilhelm Uhde, a Jew forced to hide in France during World War II, put it: "His story is not that of an eye, a palette, a brush, but the tale of a lonely heart which beats within the walls of a dark prison, desiring and suffering without knowing why. Until one day it saw the sun, and in the sun recognized the secret of life. It flew towards it and was consumed in its rays." No wonder, I think, that no modern artist is more beloved.

And as with painting, so with music. You can make a very long list of popular songs about the sun—Stevie Wonder, Bill Withers, James Taylor, Elton John, yes, of course, John Denver and his shoulders. Obviously the Beach Boys. Sheryl Crow, Katrina and the Waves, Natasha Bedingfield, Weezer, "You Are My Sunshine." Fats Waller may have written "On the Sunny Side of the Street"; Louis Armstrong and Billie Holiday and Frank Sinatra and a thousand others sang it. Heck, Sun Records produced Elvis and Johnny Cash. But in my lifetime the most popular musicians of all were the Beatles—their catalogue runs impossibly deep; every person reading these words could recognize 50 of their songs from the opening bars. But with all the paeans that Lennon and McCartney deserve, the band's most popular song (by far) is the George Harrison composition from which this book takes its title. It's been streamed 1.6 billion times on Spotify, twice as many as "Hey Jude" or "Let It Be" or "Yesterday." I think it's because we live at a hard moment and because it points the way toward something better.

So now let's talk about the orb itself, the object of this veneration, the thing that provides us with light and warmth and, via photosynthesis, our food, and which now is willing to give us all the power we could ever use. It's the one important object that we can't really look at. So what actually is the sun?

It took Western science quite awhile to come up with ideas half as good as "flying emu egg." Clever Babylonians calculated the solar year at just over 365 days, and clever Greeks (credit goes to the polymath Eratosthenes) calculated the earth's circumference using the sun's position in the sky. The Indian scientist Aryabhata figured out from the apparent movement of the sun that the earth was rotating on its axis, and at the height of the Islamic scientific flowering, the mathematician and astronomer Ibn al-Shatir not only figured out how to use sundials to create equal hours across the seasons of the year but also anticipated the Copernican idea that the sun and not the earth might be at the center of it all. Galileo used an early telescope to observe the movement of sunspots, providing the first evidence that the sun rotates. Isaac Newton, who at 22 temporarily blinded himself looking at the sun, showed that its light broke into a spectrum of colors. But what *was* the sun?

The question became, if anything, harder to answer as time went on—mostly since scientists began to think that the sun must be very old indeed. That's because—aided by clues like the layers of rock on earth, and the fossils embedded in them—they began to suspect that the earth was very much older than the 6,000 years that the Bible suggested, and it stood to reason that if the earth had been around for a long time, the sun probably had, as well. But what could keep burning up there for all that time? A big ball of coal seemed unlikely—humans knew enough about coal to understand it would have long since burned out. Something had to be renewing the fire—for a while, one prominent theory held that meteors and asteroids were being constantly pulled into the sun's gravity, and in the process generating great quantities of heat. But that, scientists eventually decided, would have required an improbable number of meteors. By the beginning of the 20th century, measuring from the radioactive decay of the elements, we knew the earth was something like five billion years old—what could have kept the fire burning that long?

I was sitting in a pleasant sidewalk café in my little town of Middlebury, Vermont, when my colleague at the college, physicist Rich Wolfson, finished up the story of the sun: "In the 1930s,

Willy Fowler, Hans Bethe—they realized that it had to be nuclear fusion, that nothing else would last that long. That hydrogen was turning into helium." It took awhile, but Bethe's classic 1939 paper on energy production in stars won the 1967 Nobel Prize in physics. And once we had this basic understanding nailed down—well, as Wolfson said, "There really are no interesting issues left related to energy generation in the sun. It's understood."

So let's say, a little more precisely, what we know about the thing that makes our lives possible. Our sun is remarkably average—middle-aged and middle-sized, one of a hundred billion stars in the Milky Way and one of trillions in the universe. It contains about 99.86 percent of the mass of the solar system, which gives you some sense of its size when you think of the giant planets like Jupiter and Saturn, or even Uranus and Neptune. A million earths would comfortably nestle inside the sun. Every second, it converts 700 million tons of hydrogen into helium. The actual nuclear reactions are tiny—each one releases about 0.0000000000044 joules of energy. But there are a lot of them, so many that it's as if 90 billion hydrogen bombs were exploding each second. The temperature in the solar core reaches about 27 million degrees Fahrenheit, and the pressure is about 26.5 million gigapascals, or 250 million times as strong as the atmospheric pressure of our earth at sea level. These numbers mean very little to us—they're so extreme, so outside our normal frame of reference, as to be incomprehensible. That's true of a lot of things about our universe—it's hard to assign any real meaning to the idea that, say, the universe is 93 billion light years in diameter. But the sun, while enormous, is not *too* enormous, not for us. It provides us each year with about 720 times more energy than humans currently use. That's a conceivable number—it means that we can harvest a small fraction of it and have far more than we need. It means that we can count on the sun.

The thing that interests me most about the sun is this steadiness, this stability. The sun rises every morning, and it shines at more or less the same intensity each day. But I do need to detour for a moment to say that's *not* what most interests solar astronomers, at least at the moment. They're fascinated by the small differences in solar output—the way that the sunspots appear and disappear, the fact that the sun can eject fountains of plasma as its magnetic fields shift and twist. Humans have known about sunspots at least since the Shang dynasty of China 3,500 years ago; we now understand that they are sites of intense magnetic activity, which cools them. More recently we've come to understand that those magnetic fields are producing intense flows of particles off the sun's surface.

Eugene Parker, who figured that out, is one of those figures who testify to the brilliance of America's 20th-century scientific enterprise, built on the back of land grant universities, elite colleges, and government labs. He got his bachelors in physics from Michigan State in 1948, and went on to Caltech for his PhD, before arriving at the University of Chicago in 1955, where he would spend the rest of his career. He was a young assistant professor there in 1957 when he calculated that the temperature of the sun's corona implied that flow of particles—a "solar wind." It was an idea immediately ridiculed; the first peer reviewer on his paper wrote, "I would suggest that Parker go to the library and read up on the subject before he tries to write a paper about it, because this is utter nonsense." But the editor of the relevant journal published his paper anyway, because he couldn't find a flaw in Parker's math, and in 1962 NASA's Mariner 2 spacecraft, en route to Venus, encountered . . . a constant stream of particles emanating from the sun.

The earth's magnetic field forms a protective bubble around our planet, deflecting this solar wind; when the sun flares and it blows at its strongest, however, it can cause interesting commotion here on earth. The solar wind is responsible for the aurora borealis, for instance—and since the mid-19th century we've suspected that can be troublesome as well as beautiful. In late August and early September 1859, an outbreak of the "northern lights" could be seen as far south as Panama, Rome, and Havana. Kathryn Schulz described this episode in *The New Yorker*: "Gold miners in the Rocky Mountains woke up at night and began making breakfast, and disoriented birds greeted the nonexistent morning." But the planet's new telegraph systems went haywire too: "At some telegraph stations, operators found that they could disconnect their batteries and send messages via the ambient current, as if the Earth itself had become an instant-messaging system." In the intervening 160 years, that pattern has repeated itself, with increasing consequences as telegraphs gave way to more sophisticated electronics—in 1967, for instance, the Air Force decided that the Soviets were jamming our early warning radars and started scrambling the bombers, only to discover that it was a solar flare. So it pays to understand—and perhaps to learn to forecast—this solar weather, which is why NASA launched the Parker Solar Probe in 2018 (Parker himself was still alive to see the launch, at age 91, the only American to witness something named for him sent into space).

On Christmas Eve of 2024, just as I began to write these pages, that Parker probe had its closest approach to the sun, just 3.8 million miles above the solar surface. It was traveling faster than any man-made object in history (430,000 miles per hour, or about 250 times faster than a bullet from a gun), so its close encounter was brief. (Necessarily brief, since it was very hot—normal cables would have melted, so a team of researchers grew sapphire crystal tubes in which to suspend the wires.) But the encounter was long enough, researchers believe, to tell us plenty. Dr.

Nicola Fox is NASA's head of science, and she explained to me as best she could the hopes for the spacecraft a few months before its Christmas flyby. "There's a region—we call it the transition region—where the magic happens. The magnetic field rotates with the sun but in this transition region the plasma itself gets very energized—superheated—and at that point the magnetic field breaks away from the pole of the star and gets swept out with the solar wind. And we can't tell what physical processes are happening in this region. Standing downstream you can watch it flowing, see what's in the water, but you can't see around the bend—what's powering it? A waterfall?" So now we'll know, or at least we'll know more. And this is both useful (a solar flare might, God forbid, crash the internet and then what would we do?) and, like all true science, beautiful.

But again, I don't want even the aurora to obscure the deepest beauty, and most important meaning, of the sun. Its greatest virtue is its incredible regularity; it's our metaphor for stability. We don't want to look at our star and imagine danger. We need to see it as our most important, most beautiful, connection.

I mean this first in very practical ways. Light from the sun travels—hey!—at the speed of light, which is 299,792 kilometers a second, meaning it takes it about 8 minutes and 20 seconds to reach us. And it reaches us in the correct quantities—obviously, if the sun was just a tad bigger, all our nice plastic would melt; somewhat smaller and we'd live on an ice ball. As it is, the sun powers just about everything: It evaporates water from the oceans, and allows plants to release water vapor into the atmosphere; it drives the winds that in turn distribute that water around the globe. The reason you might get sad in the winter is because sunlight increases the brain's release of serotonin; exposure to its ultraviolet rays can produce endorphins that make you feel sweet. (A 2012 study found that pregnant women who were in their third trimester during sunny months gave birth to children who were less likely to develop depression later in life.) Expose your skin to sunlight and your body produces vitamin D; your immune system may strengthen. Even the stuff that's harmful is good—our atmosphere and magnetosphere blocks most of the sun's harmful ultraviolet rays and energetic particles, but enough leak through to cause the occasional genetic mutations that power evolution. (This is why going to Mars is going to be hard—without sufficient atmosphere to block the sun's stream of particles, human tissues, including the brain, get hammered.) None of this is surprising; since we evolved on a planet with a sun, that sun is necessary for us. Almost by definition, the amount of sun we get is the right amount.

Or at least it was until we started screwing around. When we poured chlorofluorocarbons into the atmosphere, that eroded the ozone layer and let more ultraviolet radiation reach the surface—people began to get skin cancer in unprecedented numbers. If we hadn't signed (and more or less followed) the 1987 Montreal Protocol and slowed the production of these chemicals, then by the middle of this century we would have been on a very stricken planet—five minutes of exposure to sunlight at the latitude of Washington would have given you a nasty burn.

And now of course we're altering the earth's energy balance, trapping more of its heat by pouring a different group of chemicals into the air. That's all climate change is—the rapid alteration of the earth's energy balance by chemical means. The carbon and methane we've poured into the air traps heat—call it an extra watt per square meter of the earth's surface. Which doesn't sound like much: that's one small white Christmas tree light. But there are a lot of square meters, and all together it's the heat equivalent of five Hiroshima-sized atom bombs every second. No mushroom clouds—just melted icecaps, rising seas, raging fires.

There really are only two ways to deal with this energy imbalance. One is to blot out more of the sun. This is what some people call geoengineering—the introduction of some new chemical, sulfur probably, into the atmosphere that would block some of the sun's incoming solar radiation. This is a bad idea—the computer models predict it would do weird things to weather patterns, not to mention all that stuff about serotonin, and also plants really like sunlight. (It perhaps won't surprise you that the biggest backers of geoengineering are what Bloomberg called in 2024 the "Silicon Valley elite.") But it's the kind of bad idea that we'll probably turn to if we don't take the other course—the one I've been outlining in this book—which is to stop burning the things that produce those sun-trapping chemicals, and instead power ourselves directly with . . . the sun. That the sun is the easiest way out of the fix we've gotten ourselves into with the sun is one of those facts that suggest God has a sense of humor.

And a sense of beauty, because we've evolved to love the sun. We mostly ignore it now, as I've said, staring instead at the small light in our palms. But sometimes we're reminded of something more primal. The year 2024 was, by my estimate, a pretty dour year—the hottest in human history, with repeated disasters including the devastating electoral hurricane in November. But for much of America there was one day of unmitigated pleasure, and that was when, in early April, a solar eclipse crossed most of the country. From the Mexican border in Texas, to the Canadian border in Maine (and indeed on the far side of both borders, since eclipses cross them with ease!), the sun went out for a brief minute, and everyone watched with wonder and delight. Cars poured into northern Vermont, because it had a nice long stretch of totality—definitely our biggest traffic jam of the year, maybe ever. But everyone was in a good mood—I watched it from the main quad at Middlebury College, and for a number of minutes a thousand college students Put Down Their Phones. Across Lake Champlain, in New York's vast upstate complex of prisons, inmates sued for the right to watch the sun dim—they claimed it was critical to their practice of Christianity, Islam, or Santeria. It was actually an atheist who filed the successful lawsuit, however. "Moments when people of vastly different faiths converge in a single shared joy are exceedingly rare," he wrote in his filing. "The price of missing any is unknowable, but substantial by any measure." Indeed. And since solar power output drops precipitously in those moments of totality, it probably makes sense to turn eclipses into legal holidays that we celebrate simply for the pagan pleasure of looking at the one thing that's forbidden us at all other times. It would be a sweet way to acknowledge all the work the sun does for us.

Forty years ago, on a long global book tour for *The End of Nature*, I found myself being interviewed by a Dutch journalist. After a lot of the usual questions about global warming, he asked me one more thing: What was my earliest childhood memory? He asked everyone he interviewed this question, he said, and invariably it was either a moment of joy or painful embarrassment. Mine, I told him, was the latter: As a young boy I lived in the Los Angeles suburb of Altadena, and one day, in shorts, I eagerly raced out to the swing set that was the fixture of so many suburban 1960s backyards. But it was a summer afternoon, and the sun had been heating the metal seat all day, and that seat burned my thighs—I felt the pain of the burn, and also of the humiliation; I'd been tricked, as it were. The sun is powerful.

I've been thinking about Altadena in the weeks that I've been writing these pages, because it burned in the great Los Angeles inferno of 2025. As I've noted earlier, the house I'd lived in those 60 years ago was consumed, after a year that featured the least rainfall and the highest temperatures ever recorded for the region. As I thought about that place, which my family left for the East when I was five, more memories kept coming back. Most were happy, especially of the first hikes I ever took in my life, which were up the fire road that led to the observatory atop Mount Wilson. In a lifetime of climbing mountains, this was the first time I ever had that sense of the world spilling out beneath me. That observatory almost burned in the LA fires too, but a skeleton crew was able to keep it safe, and thank . . . heaven. It was up there that the great astronomer Edwin Hubble, using the 100-inch telescope that was then the biggest in the world, first figured out that our Milky Way was but one small galaxy in a vast universe—and that that universe was expanding, setting the ground for our understanding of the big bang. But the telescopes on Mount Wilson also explored our own solar system, and in 1962 (while I was living below) an astronomer named Robert Leighton discovered oscillations all across the solar surface. Measurement of these waves allowed others to model the solar core and to calculate the rate at which it consumes hydrogen. These are the numbers that assure us the sun will be around another five billion years or so.

Which is long enough for me. Someone else can worry about what we'll eventually do when it turns into a black dwarf. For now it provides all we need.

When I wrote the last words of *The End of Nature*, I was very nearly catatonic. Wrestling with the import of global warming, especially at a moment when there were very few others to share my angst, had almost done me in. If we were henceforth to inhabit a more hostile planet where every heatwave and every flood reflected our own folly, I wrote, then for philosophical comfort we'd have to look up at the great expanse of stars, beyond our reach, which still held mystery and wonder. Many a night since, under the dark mountain skies where I live, I've stared up at the Milky Way and felt that small relief.

I end this book saddened, too, of course—saddened by all that has happened in the last 40 years, and by all that we haven't done. But I also end it exhilarated. Convinced that we've been given one last chance. Not to stop global warming (too late for that) but perhaps to stop it short of the place where it makes civilization impossible. And a chance to restart that civilization on saner ground, once we've extinguished the fires that now both power and threaten it.

If we're to do that, we'll have to turn to the *daytime* sky. And to one star in particular. Our star.

Notes on Sources and Acknowledgments

I've tried to credit sources in the text, and since much of this book covers events of the last two or three years those sources tend to be journalistic. *The New York Times* and *Los Angeles Times*, *The Washington Post*, and *The Wall Street Journal* (if you avoid the editorial page) have dramatically improved their climate coverage in recent years; since this revolution is happening fastest in other places, the *Financial Times, The Guardian*, and *The Economist* often have some of the best reporting. You can find the trade press on the web, but the go-to source of sources is Bloomberg, and especially its New Energy Finance team—if there's one supreme chronicler of this transition it's BNEF's Jenny Chase. I'm also grateful for podcasters like David Roberts, and for outfits like Inside Climate News, Canary Media, Heatmap, and Grist. When I started work on this beat back in the 1980s there were almost no others engaged in the work; no one could be happier for the company.

I've tried to describe Mark Jacobson and Kingsmill Bond in these pages, but they are two crucial figures in this moment. My

Middlebury colleague Rich Wolfson absorbed many questions about solar physics and about our energy system; he remains the best college lecturer on climate change I've ever seen. Since I knew virtually nothing about the sun when I began, I'm indebted to books by people including Michael Carlowicz, Keith Barnham, Sunil Amrith, Philip Judge, Bob Berman, Leon Golub, Jay Pasachoff, Ryan French, and Richard Cohen. Dr. Nicola Fox, head of NASA's solar science programs, was kind enough to talk with me at length even amid the most ambitious year for solar exploration in human history. Jigar Shah at the Department of Energy, Billy Parish, and Mary Powell from the solar industry—everyone's generosity has been deep and welcome.

I'm grateful for the community that has sprung up around my newsletter *The Crucial Years*—readers are often the source of excellent tips. (And if you want to keep up to the moment on these topics, it's not a bad stop.) But it was really *The New Yorker* that allowed a great deal of the reporting in this book—I'm deeply indebted to my editor there, Virginia Cannon (there are long sections of this book that were edited first by her steady hand) and to the factcheckers and other staff all the way up to David Remnick, who has kept that magazine flourishing against all odds. My other institutional home is Middlebury College, where I owe huge thanks to Janet Wiseman, and to wise leaders, above all Laurie Patton and now Ian Baucom.

For me, writing and activism are often intertwined. My list of colleagues and friends in the organizing world is endless, but I'd like to pay particular thanks to the people who I've worked beside to build Third Act over the last four years—they include Vanessa Arcara, Kafia Ahmed, B Fulkerson, Anna Goldstein, Deborah Moore, Jeremy Friedman, Simone Salvo, Janina Klimas, Melanie Griffin, Pam Murphy, Christine LaRusso, Tyler Stern, Kyla

Woods, Heather Booth, Bruce Hamilton, and Lani Ritter Hall. Without the cheerful and graceful help of Veronique Graham, I'd have long since lost track of my work, and likely my mind. Akaya Windwood has been my sister in this work, and in much else. Rev. Lennox Yearwood and Rebecca Solnit have been colleagues across so many battles; the same for Naomi Klein and Terry Tempest Williams and Ayana Johnson, and all of my old friends at 350.org. And, of course, Jamie Henn, the finest organizer I know, in part because there's no one more fun to work with. On to Sun Day, with Deirdre Shelley, Duncan Meisel, and dozens of other fantastic organizers!

This is the first time I've gotten to work with W. W. Norton, the independent employee-owned publishing house that has produced so many of my favorite books. The appeal of working with legendary editor John Glusman was a big reason I ended up there, and it's been a truly congenial experience; Rebecca Springer and Michael van Mantgem pointed out in impeccably kind fashion the many places where my prose could have been clearer; special thanks to Rachel Salzman, who had to figure out the publicity campaign before I'd turned in the manuscript, and to Wickliffe Hallos, who was the chief expediter. Indeed, everyone at Norton has shown their chops by producing this book at a much-faster-than-usual pace; they understand, I think, this moment in our history.

Gloria Loomis has been my agent since my first book, back in the 1980s; not many writer-agent combinations have thrived through 20 books, and I am very much in her debt (and thanks as well to Julia Masnik, her gifted colleague).

This book is dedicated to my first grandchild, Asa Caleb Crane, who turned a year old shortly after I turned in this manuscript. He, as much as anyone, got me through the writing: Watching him start to explore his world inspired many of the thoughts that ended up in these pages. His parents, Sophie and Josh Crane, are raising him with love and panache; my wife and I watch with admiration and endless affection. That wife, Sue Halpern, is also my dearest friend; we've been together for 40 years, engaged in an endless conversation that illuminates these pages. She is my sunshine, my only sunshine.